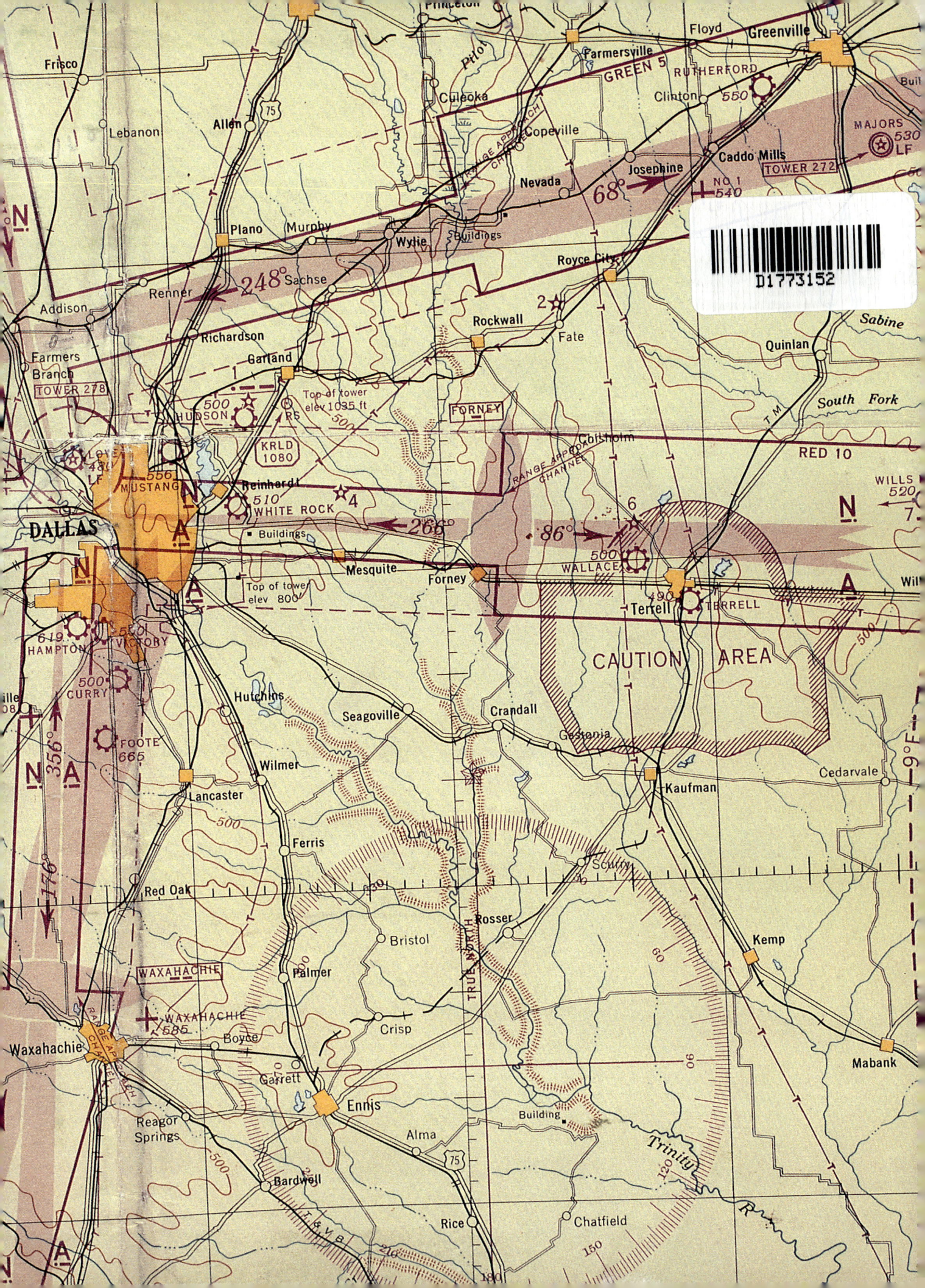

WOMEN IN PURSUIT

Flying Fighters for the
Air Transport Command Ferrying Division
during World War II

A Collection & Recollection

Kay Gott

KAY GOTT

WOMEN IN PURSUIT

FLYING FIGHTERS
for the Air Transport Command
Ferrying Division during World War II

KAY GOTT

A Collection & Recollection

```
q 940.544      55 00
Gott, Kay.
Women in pursuit
```

Copyright ©1993 by Kay Gott

All rights reserved. No part of this book may be reproduced or utilized in any form or by any means, electronic or mechanical, including photocopying, recording or by any information storage and retrieval system, without permission in writing from the publisher. Inquiries should be addressed to Kay Gott, P.O. Box 2813, McKinleyville, CA 95521.

Printed in the United States of America

Library of Congress Catalog Card Number
92-081820

ISBN 0-9633-075-0-9

Contents

Foreword 7
Acknowledgments 9
Preface 11
Prologue 13
The Beginning 17
Women Airforce Service Pilots 33
Women in Pursuit 57
2nd Ferrying Group 95
3rd Ferrying Group 119
5th Ferrying Group 139
6th Ferrying Group 167
21st Ferrying Group 197
Epilogue 209
Appendices 217
References 221
Index 225

✯✯✯

*This document is dedicated
to Myra Gott, my Mother,
who is my best fan
in my flying, and my life.*

✯✯✯

Foreword

In the fall of 1990, I had been asked to deploy to the Gulf War as the aircraft commander of a KC-135 stratotanker and I was anxious to go. At last I would do what I had trained all my military career to do, go to war. A week before my departure date I flew to Seattle, Washington. As I stood before the distinguished audience of women it was difficult to find the words to express my profound admiration and respect. There were so many thoughts that went through my mind, so many friendships I had made, the mentors who were always there to listen and advise me. Now it was my turn to thank them for giving me and my sister aviators the opportunity to serve our country. Just as they, the Women Airforce Service Pilots (WASP) had served in World War II.

In 1942, before I was even a gleam in my parents eyes, an exceptional group of women laid the foundation for the future of women in military aviation. Coming from varied social, economic, and educational backgrounds, they were indeed "the girl next door." The one thing they shared in common was their love of flying and dedication to their country. With this in mind they willingly took on the task of flying any and every aircraft in the inventory. From the tough little T-6 trainer, the hot P-51 Mustang and the big iron bombers and transports the WASP flew them all. Although the Women Airforce Service Pilots existed for a relatively short period of time in our history, 1942 - 1944, these women of character, courage and dedication made a mark that will never be forgotten.

Over the past twelve years I have had the pleasure of meeting many of the WASP. These women are mentors in the truest sense as they continue to actively support the advancement of women in military aviation. As a result of their encouragement, their legacy formed its own organization Women Military Aviators (WMA). Between the WASP conventions and the WMA conventions, two generations of women military pilots have the opportunity to meet each year and share aviation experiences. The more I learn about the WASP the more I admire them. Each woman has her own unique stories, perspectives, and humor to share.

Today we are preparing to celebrate the fiftieth anniversary of the Women Airforce Service Pilots. What better way to honor them than to invite you to meet them "in person." Kay Gott, a member of the class of 43-W-2, gives you the opportunity to turn back the pages of time and join her and her classmates in their WASP experience. Through her pen, dreams are born again, aspirations are realized, moments of happiness and sadness are captured. Join in the camaraderie of the WASP and the spirit of adventure that belongs to those who live history in the making.

Kelly S. Hamilton, Lt. Colonel
U.S. Air Force

Acknowledgments

My husband Keith, who was patient and helpful through the months and years of this research and writing.

All the WASP who responded to my pleas for those old photos, consented to open their logbooks and footlockers, and gave me enthusiastic interviews.

Philip Dresser who copied 50-year-old photographs and made them look like new.

Lindsay Olsen for additional photography work.

Victor Jacoby, who organized data when I was floundering.

Carolyn Heilbrun whose book *Writing Women's Lives* led to an examination of the human side of this particular experience that was ferrying fighter aircraft.

Patsy and Peggy Givins who advised me to zero in on one particular WASP incident at a Pro-Pen writers' retreat weekend at Kimtu.

Laura Montagna, Creative Type, typesetting and publishing, and Margo Coleman, printing services specialist, who copy-edited and proofread; both saw my work through to completion.

Savor this moment
              this time
In our lives.
            Treasure it --
It will never
           Come again.

                  -Kay-

Preface

We were a pretty remarkable bunch of women. We were hand-picked and we had high qualifications.

Each woman pilot was unique. Each flight was unique. The entire WASP experience was treasured and carefully logged at the time it occurred.

★★★

I felt that surely, in the Smithsonian, at our Nation's capital, there would be records of women who flew. I was just looking for records of women in aviation generally. I was so appalled at the paucity of information on **any** woman. One picture of Jacquline Cochran, and one picture of Amelia Earhart; a beautiful sculptured bust of Amelia Earhart. And then the misinformation of Nancy Love's photograph being mislabeled as Amy Mollison, a British woman pilot.

It has been impossible to find one source of information about our history. It has taken five years of intensive work to locate all the WASP who ferried pursuit for the Air Transport Command (ATC). Some women in the Training Command flew fighters, but that is another story. The WASP who flew pursuits have been most generous to loan their treasured logbooks, photographs and pursuit orders.

The WASP were too busy to keep scores. They did not operate on the "star" system, nor did they have "aces."

Experience in the WASP was not a competitive venture. There was no time to contemplate or to know what the other WASP were doing. Even thirty years later, reading the interviews with three WASP at a reunion, they inquired from each other knowledge of WASP history. In 1972, WASP knew little of the accomplishments of the group of women pilots — 1102 in number, and who did a great variety of flying tasks, not just ferrying pursuits. In fact, by 1972, WASP history had faded into the woodwork.

Out of the 1102 women who flew military aircraft for the United States government in the years of 1942 to 1944 in World War II, less than one-fourth of them were attached to the Air Transport Command. The other three-fourths were pilots for the Air Training Command, who flew a great variety of types of missions, and in a great variety of aircraft. Their stories are told elsewhere.

This document is specific to the small number, only 134 women, who flew fighter aircraft during World War II for the Air Transport Command. Their mission was transporting pursuit aircraft within the United States to where these airplanes were needed.

★★★

What did it mean to be ferrying fighters?

Our job ... our mission ... our task was to get the airplane to its destination in the quickest possible time, in the best possible shape.

Sometimes it meant taking a brand-new airplane right off the assembly line, checking it on the ground, then test-flying it. After we made sure it was ready, we flew it to where it was needed.

There was a war on, and someone needed that airplane if we were to win that war. Women pilots were called to serve their country.

We did serve, and we served our country well.

Dr. Dorothy Schaffter wrote that the WWII experience produced permanent gains in the status, freedom and capacities of American women — women's move toward self-reliance. Though the use of women pilots in the military was truly experimental in World War II, she wrote in 1948 that "It will never again seem revolutionary for women to fly fast pursuit planes."

The fact that WASP flew **everything in the air** from 1942 to 1944, not just pursuits, seems forgotten. Twenty-six years were to elapse before women were allowed to fly military aircraft again in the United States, in 1970.

An airplane just demands a skilled pilot, be it male or female.

Disclaimer

This search for a bit of history began in 1987, some forty-three years after it happened. Many of the women involved are dead — memories are enhanced by the softening shades of time. This searcher attempted to get accurate documentation before we WASP were **all** gone. It continues to be a remarkable experience.

All photos and documentation are from Kay Gott's collection unless otherwise noted. 2"x2 1/2" portrait photos are from class yearbook.

Prologue

Classmates of 43-W-2 Betty Jane (BJ) Bachman, Kay Gott and Marie Muccie get together prior to a convention. (Marie flew B-25s out of Kansas City, Missouri. Kay and BJ flew pursuits out of Dallas, Texas.)

Kay: I have read Historic Document 55 enough to know it is not a satisfactory document; and if it is cited as the best document on the WASP with the Air Transport Command, it is a horrible put-down on what women did. And I agree with Jacqueline Cochran that it dwelt too much on petty controversy and not on the accomplishments of women.

★★★

Marie: We WASP have one line in the World Book Encyclopedia.

K: The miracle is, that we got to fly at all. Thirty years ago. That is another thing I find issue with the Historical Document 55. They never come out and say how we got there! We just finished a horrible depression in 1939. We scrounged up $50.00 so I could to go college. And then to FLY!

M: I didn't have a dime for flying; I worked for every bit of my flying time.

K: It is so interesting that women then flew, and women today fly in astronomical numbers. Its amazing that women, against a lot of laws, got into the air at all.

M: The WAFS under Nancy Love went in before we did. All of them had 500 hours and a commercial license. They didn't go through training when they reported to Wilmington Base. They had check rides.

K: I never realized that they didn't go through training.

M: Then Cochran wanted to start these women with training school.

K: Reading this history, I came to realize that Cochran was Director of all women pilots, and Nancy Love was the executive for the Ferrying Division. That was never clear to me before. Cochran's concept was that women would do many other things besides ferrying airplanes. Right from the start, women would be used to relieve men whenever necessary, not just in ferrying airplanes.

Cochran had a very broad concept of the use of women pilots. As Nancy Love's program was conceived, it was written right in the order that women were to fly the liaison planes; right in the authorization that they were only going to fly this type of plane. Cochran never agreed with that, right from the start! She didn't put any limit on the type of flying for women. And the early WAFS under Nancy Love flew the "L" trainer planes for a very long time. Now, there is a mysterious thing: whatever happened to the few women that Cochran took to England to fly earlier in 1942? She was recalled from England by General "HAP" Arnold to form the Women's Flying Training group.

M: Was this the ATA, Helen Richey's group? I applied for that. I got a letter back saying that I was one inch too short. I had the qualified hours; I had everything for it.

K: How tall are you?

M: Five feet two.

K: Five feet two? Do you realize that right now you are two inches short of Jacqueline Cochran's recommended 64 inches for women pilots. She recommended at the end of her document that women not be accepted less than 64 inches because of the leg length to reach the controls.

M: Right. I had to have a pillow all the time. So did Vi Thurn, Betty Gillies and Ruth Adams.

M: To get into the WASP, I went to interview with Miss Cochran in New York; her office was there. Miss Cochran said that I would have to get permission from my parents since I was not yet twenty-one. My father hit the ceiling! He was so upset. I said that if he did not give me permission, I would go and join the British women pilots; there I didn't have to be twenty-one to fly pursuits; I'll go over there. I didn't tell him I was too short. He told me he would rather have me over here, so he gave me permission. I was nineteen, almost twenty.

K: Cochran's recommendations at the end were to take younger pilots, as those as young as 18 proved better pilots than the ones over 27, yet a little limited on responsibility.

M: They took men cadets age eighteen; they were just boys.

K: Sure.

M: After they took the women with 500 hours, then they had to take women with fewer hours. There were 23 in the first group; either women had other obligations or there really were not that many women with 500 hours flying time.

M: Did those original WAFS take training?

K: They took transition, and I am able to document some of it. Helen Richey was one of the older pilots, served with the ATA in Britain, came back and graduated in class 43-5, September 11, 1943. She went first to Wilmington ferrying, and then to Kansas City, flying B-25s.

M: We took qualifying tests all the time.

K: Surely, in those tests, is some indication — were we an unusual group?

M: The men had to take the same tests.

K: Are flyers unusual people? Are they action people?

M: They have one thing in common; they like to fly. Or they wouldn't be in the darn airplanes to begin with.

K: Exactly. And they weren't all wealthy people, because you and I got in.

M: No, we weren't.

K: So they were persevering people, especially the women. Much more than the men.

M: Men didn't have to persevere, they **fell** into their training.

K: All they had to do was pass the written (and physical). All they had to do was go to a recruiting board and say "I want to fly."

K: Am I right in this? All of us women had to have some flying time? This is another question that has never been answered, and isn't even answered in the historic document.

M: That was one of the requirements when you first went in; they lowered the hours, but they never did eliminate hours.

K: Was there ever any woman who went through all our program without a prior flying time at all? I would like to know. I have always wondered.

M: Do you know the reason they felt women should have some flying time behind them:

K: No, why?

M: If a woman decides she wants to go take this pilot training, she's going to be a WASP and fly the Army airplanes, she gets in there and goes through all this expensive training. Then after she's in there she decides she doesn't really like flying, and doing all those spins and all those stalls, and quits!

K: When I went in I didn't think I was going to fly fighters. I didn't think I was going to fly twin engine.

M: I went in to fly whatever they would give me!

K: Any type of plane!

M: I always thought they would give us light airplanes nobody else wanted to fly.

K: No illusions, whatsoever. Just to fly anything.

M: And yet they put us in those brand new airplanes that had only one hour test flight. Some of them didn't have an hour. Brand new factory airplanes.

★★★

K: Did you read about the WASP in Flying magazine?

BJ: Yes, someone brought it to us.

K: Do you feel that the WASP, the women are documented.

BJ: No. "The Stars at Noon," Jacqueline Cochran's autobiography, has a small mention of the WASP toward the end of the book.

K: Do you know where you could find anything about the WASP? Have you ever tried to look it up in the library? Or anyplace else?

BJ: No, I haven't. In fact, no one knows about them. And people say "Oh, I hear you used to fly," and then when you mention the organization, they don't know what WASP was.

M: The only ones who would know are the ones who were in World War II, like the guys; the Korean War, they don't know who WASP are — younger group, anyway. We are just not known.

K: How would it benefit us to be known now? We are not getting any benefits.

BJ: I think we should have the same rights as the GIs. I don't mean money. For example, if we are alone and ill, we could use Veterans' hospitals.

K: Recognition might make us **feel** better; but the significant thing I think we could do, is make it possible for women **now** to fly. Why should we have to have a **war** to give women the opportunity to fly? Women shouldn't have to have a war.

I think this is one of the contributions we could do; compile a bibliography on what's written about women in flying.

The Beginning

If you were an American woman and longed to fly, the best place to head for in 1939 was to a college lucky enough to get into the government aviation program.

Can you imagine a program in aviation having no publicity, buying no airplanes, renting no classrooms, hiring no instructors? And teaching flying? This was such a secretive experiment you almost had to know somebody who knew how to get in. Only one woman was permitted for each ten men — so even then, your chances of entering the program were slim, for who wanted to have a female in an all-male class?

The Depression was still on. There were few jobs, and little cash for flying lessons. One farmer's daughter went to the College of Idaho on the load of celery her father delivered to the dormitory; the students ate celery all winter from her "tuition" payment. Times were tough!

Italy, France, England and Germany had many more flyers than the United States. The U.S., however, saw Lindbergh fly the Atlantic twelve years earlier, and Amelia Earhart lost in the Pacific for two years; mail was now flown across the continent; and barnstorming air circuses were popular.

It was the vision of one man to send the youth of America into the skies in 1939.

It took this one man, Robert H. Hinckley (1892-1988), a member of the Civil Aeronautics Authority, to lay the plan before a receptive President Franklin D. Roosevelt. A group of far-seeing committee members found the financing and the way to bring aviation to the students of America. In 1983, the Federal Aviation Administration honored Robert H. Hinckley, founder of the Civilian Pilot Training program that trained fliers just before the United States entered World War II, saying that this college-based program helped shorten the war by giving the U.S. thousands of trained pilots.

OBITUARIES

Robert H. Hinckley

Eden, Utah

Robert H. Hinckley, founder of the Civilian Pilot Training Program that trained thousands of fliers just before the United States entered World War II, died of natural causes on Saturday at the age of 96.

In 1983, the Federal Aviation Administration honored Hinckley for developing the college-based program, saying that it helped shorten World War II by giving the United States thousands of trained pilots and airmen.

During the Roosevelt administration, Hinckley was a member and later chairman of the Civil Aeronautics Authority, forerunner of the FAA.

After his retirement, Hinckley established the Hinckley Institute of Politics at the University of Utah, which brings U.S. and foreign political leaders and administrators to the university to instruct political science students.

United Press International

5/22/88

Art Mortenson, Kay Gott and Instructor, Harry Clark with the Porterfield training plane.

In the 1940s, more CTP pilots learned to fly in the No. 1 Putt-Putt, Piper Aircraft's J-3 Cub, than any other small plane. (Photo courtesy of Robert F. O'Neil)

Civilian Pilot Training Program

The CPT, Civilian Pilot Training (CPT) program, began as an experiment in 1939 using colleges and college teachers for the ground school classroom teaching, and the airplanes and flight instructors from nearby flight schools that formerly had languished for students. Eventually, a CPT program was initiated in every state and Puerto Rico.

By June 1941, more than 2000 newly trained, young women pilots were scattered throughout the land. Many of the women pilots gained the required hours to enter the WASP program while in the Civilian Pilot Training program.

The first and foremost purpose of the CPT was the upgrading of civilian aviation. But there were many spinoffs to this program: aeronautical courses in the colleges; academic credit for ground school work; new flying fields; small operators becoming responsible business men; and flying opportunities brought into hundreds of communities and to thousands of families.

The CPT program terminated for women in June 1941.

Author's note: Astronaut John Glen is perhaps CPT's most renowned graduate.

College of Idaho Civilian Pilot Training Class, Fall 1940. Back row, Leland Thomas, Franklin Turmes, Ronald Blakely, Earl Howe, Bill Hunt, John Mather, Louis Koutnik, Gene Poppaw, Mason Brown, Gordon Harbert, Kay Gott, Ed Dolton, Amos Roden. Front row, Theron Gough, Bill Reed, Leo Patrick, Wayne Rife, Air Instructor George Cook, Air Instructor Harry Clark, Ground Instructor Steunenberg, Bob Jones, Bill Ward, and the aircraft owner and his child.

College of Idaho 1942 Flying Club. Back row, Carl Gustafson, John Young, Jim Hoagland, Jim Harrah, A. Roberts, Bill Sanders. Middle row, Fred Greenfield, T. Stearns, Kay Gott, Layton Patterson, Stanley Newman. Front row, Paul Oakes, Louis Koutnik, Theron Gough, G. Chester, Ernest Skinner, Norman Maffit.

College of Idaho Flying Club, organized Fall 1939. 3rd CPT class began November 1940 and finished March 1941 with 15 students. Top row, Bruce Rivett, Warren Roberts, Ronald Blakely, Marjorie Tucker, Kay Gott, Howard Games, Carl Dunaway, Leo Patrick. Middle row, Ernest Clapp, Roger Cartwright, Gerald Ruddell, Pete Dustman, Franklin Turmes, Tad Hankins, Gerald Graves. Bottom row, Jim Harrah, Gene Poppaw, Cliff Starns, Joe Coon, Merle Armstrong, Lee Thomas.

War Training Program

When the Civilian Pilot Training program ended, it was replaced by the War Training Program (WTP), and cadets in military uniforms invaded the colleges. The objectives of the CPT and the WTP were drastically different. The CPT program was designed to turn out pilots with skill in the air and a private pilot's license, mainly upgrading civilian aviation. The WTP was military-directed with the sole objective to win the war.

Dr. William W. Hall, President of the College of Idaho, Caldwell, Idaho, wrote of those WTP times: "The college contracted with the Army Air Forces to furnish instruction in English, mathematics, history, physics and geography; to provide physical training and courses in medical aid and civil air regulation. Military drill, discipline and instruction in military customs and practice were in the hands of the assigned military personnel." Dr. Hall continued: "The War Training Program, from parts of speech to mathematical formulae, was consciously related to a stern and inevitable alternative, namely, to kill or be killed."

The War Training Program ended abruptly in July 1944 leaving the civilian flight instructors without jobs and eligible to be drafted into the "walking army" since it was deemed there was a sufficiency of pilots. It was these civilian flight instructors who lobbied Congress in 1944 stating that the women in the air — Women Airforce Service Pilots (WASP) — were taking their jobs.

Author's Note: Myra Gott taught both CPT and WTP students.

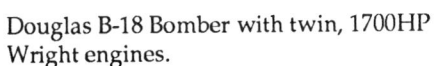

Civilian Pilot Trainees at the College of Idaho, Caldwell, visit a B-18 Bomber that landed at Caldwell Airport. Ground School Instructor Myra Gott is the woman in the photo. Jack Hoke is at the far left. 5th from left is Joe Dion, source for this photo, who dated the event circa March 1943, class 43-E, CPT. The others are unidentified as of publication.

Douglas B-18 Bomber with twin, 1700HP Wright engines.

HEADQUARTERS
3060TH AAF BASE UNIT (COLLEGE TRAINING, AIRCREW)
COLLEGE OF IDAHO
CALDWELL, IDAHO

30 June 1944

Mrs. Myra Gott,
1019 12th Ave., S.,
Nampa, Idaho

Dear Mrs. Gott,

    This is to express appreciation for valuable services rendered as a member of the academic staff of the College of Idaho participating in the aircrew college training program. Your personal interest in the aviation students and your punctual attendance at classes in spite of inadequate transportation to and from your home reflected a spirit of cooperation especially commendable.

                                  Respectfully,

                                  LAWRENCE J. HENRY,
                                  1st Lt., Air Corps,
                                  Commanding.

LJH/rgr.

Letters received by Myra Gott when the War Training Program ended at the College of Idaho.

The College of Idaho
ESTABLISHED 1891 BY
WILLIAM JUDSON BOONE
Caldwell, Idaho

WILLIAM W. HALL, JR.
PRESIDENT

July 10, 1944

Mrs. Myra Gott
1019 12th Avenue South
Nampa, Idaho

Dear Mrs. Gott:

Just a word of appreciation for your part in the aviation program including the C.A.A. training. We are sorry that the program of the Air Forces terminated so abruptly and that we could not have had you with us longer. We shall miss you greatly.

Please extend our best wishes to Kay when you write or see her.

                              Cordially yours

                              William W. Hall, Jr.

WWH:ms

Air Transport Auxiliary

In October 1938 under Civil Air Guard, Britain had subsidized flying lessons open to both sexes, ages 18 to 50, who passed the private pilot's medical. Some 3,000 to 4,000 men and women had obtained their license by the time war was declared on September 3, 1939.

When Hitler marched into Poland on September 1, 1939, Great Britain and France declared war on Germany two days later. As Germany advanced through the Netherlands, Belgium and France, a great armada of 900 vessels gathered in England to evacuate troops from Dunkirk to safety across the English channel, May 26 to June 4, 1940. Nazi bombing of England began July 20th. As the Royal Air Force (RAF) attacked German aircraft and stopped the threat of invasion, Prime Minister Churchill said: "Never in the field of human conflict was so much owed by so many to so few."

At the time of Dunkirk, the RAF pilots were tied up clearing aircraft from factories, since factories were prime targets for enemy bombing. Because of this shortage of pilots at the Dunkirk event, it was evident that there was a use for women pilots. Eight women who were flight instructors were chosen for the Air Transport Auxiliary (ATA), given short-term contracts, and spent the first years ferrying light aircraft, restricted to the "Tiger Moths." Miss Pauline Gower, commander of the women pilots, was authorized to pick five women to learn to fly larger aircraft.

During 1940, the number of women pilots in the ATA increased from the eight who started at Hatfield, England, to a maximum of 110 during ATA's five years. As the women pilots' competencies in flying grew, so did their transition to heavier aircraft.

★★★

Some 200 American male pilots signed contracts with ATA between August 1940 and September 1941; but most left the ATA for home when the United States got

Women Pilots of the Air Transport Auxiliary, England, stroll past a Spitfire. Roberta Sandoz Leveaux and Kay Van Doozer, USA; Jalwiga Pilsadski, Poland; Mary Hooper, USA; White Watham Airfield, England, 1943, (Roberta Sandoz Leveaux collection)

Air Transport Auxiliary Women Pilots braved the England elements not only flying but also posing for this photo with a Spitfire fighter.

into the war after Japan attacked Pearl Harbor on December 7, 1941.

The men and women of the ATA came from a total of 28 countries.

✯✯✯

In Britain, the women pilots might fly bombers one time, and fighters on the next assignment. The British ATA pilots carried "Pilots Handling Notes" with them on each airplane. These notes were specifics on such things as take-off speed and throttle settings, and were a good check list of each plane.

Joan Hughes became the only female instructor for both sexes on all classes of aircraft including four-engine bombers. Lettice Curtis, in her five years with the ATA, delivered 400 four engine aircraft, 150 Mosquito airplanes, and over 50 different types of airplanes.

✯✯✯

Jacqueline Cochran and Nancy Harkness Love, two outstanding American pilots, were both having difficulty persuading United States leaders that there was untapped women pilot power in America waiting to be called. Jacqueline went to England on January 24, 1942 to confer with British Commander Pauline Gower. In the spring of 1942, she took 22 American women pilots to join the ATA.

Recruited by Miss Cochran, they met in Montreal for one, tough flight-check, and were put on convoy ships for the hazardous, zig-zag ocean voyage through the German submarine wolf packs. Roberta Sandoz and Emily Chapin were on the same ship, became fast friends, and recounted they were very seasick on the two-week trip in the Norwegian freighter Mosdale out of St. Johns, New Brunswick. Roberta recalls, "Convoy crossings were agonizingly slow and nerve-wracking, losses from the perimeter being not uncommon."

Later, Roberta had a wild adventure in a Spitfire fighter, but successfully brought it down, only to be called before a Court of Inquiry for the incident. The report reads: "23 February, 1944. The aircraft force landed without damage at the airfield. Cause: The pilot was unable to reduce engine revolutions below 3,000. Technical investigation showed a defect to the propeller hub mechanism. Responsibility: The pilot is held not responsible for this forced landing." This is a typically laconic English summary of miraculous flying skill.

Roberta writes of the ATA experience: "We pre-WASP held our own in England, and demonstrated day after day in rotten weather that women could indeed handle military aircraft safely and efficiently."

✯✯✯

When their 18-month contract was up, five of 22 American women pilots returned to the United States to enlist in the Women Airforce Service Pilot program (WASP) training class:

Helen Richey, class 43-W-5
Hazel Raines, class 44-W-3
Myrtle Allen, class 44-W-8
Emily Chapin, class 44-W-10

Katherine Van Doozer was over the age limit of 35, and served as Staff Advisor for the Women's Flying Training Detachment at Avenger Field, Sweetwater, Texas.

✯✯✯

Jacqueline Cochran wrote: "Every type of plane was flown by this group of American women pilots. They would move planes from one base to another or from factory to field or vice-versa. Their accomplishments were largely unsung even when compared to the little-known work of the WASP at home, but they were noteworthy."

"Flying the corridors set off and hemmed in by captive balloons was not easy, particularly in sticky weather. The same could be said for flying planes in need of repair. Mary Webb Nicholson, my secretary at home, was the only American to lose her life. The propeller flew off her plane and she crashed in rough terrain."

"Only one girl proved unsatisfactory to the British. She was really one of the best pilots of the lot, but through sickness at home or because of other emotional problems of her own, she failed to do her best and eventually cracked up a couple of planes. One time on a take-off she hit a hangar roof and in the collision lost a wing tip and then, while steadying her plane from this she hit a telephone post beyond the hangar and damaged the other wing. Notwithstanding all this she still was able to circle the field and make a safe landing. Later on when I was in charge of women pilots' training with our own air forces at Fort Worth, Texas, and this girl had been released from English service, I had her given a refresher course at our training base at Sweetwater, Texas, then put her at work with the WASP. She served until the WASP were deactivated."

Helen Richey, WASP, class 43-W-5, in her ATA uniform which she wore when serving in England before the WASP were formed. (Photo from the collection of Marie Muccie who served as Helen's co-pilot in a B-25 out of Kansas City, Kansas, 33rd Ferrying Group, U. S. Air Transport Command)

December 1942 WAFS Katherine Rawls Thompson, Phyllis Burchfield, Nancy Batson, Delphine Bohn, Florene Miller and Teresa James with primary trainer bi-planes in background. (Florene Miller collection)

Helen McGilvery, Teresa James, Dorothy Fulton, Fairchild Aircraft representative, Kathryn "Sis" Bernheim, Gert Meserve, Betty Gillies and Nancy Batson in front of a PT-19.

Women's Auxiliary Ferrying Squadron

Nancy Harkness Love began working a civilian administrative job with a ferrying division of the Air Transport Command (ATC). By chance, Colonel William Tunner, head of the Army Air Force Ferrying Division Domestic Wing, heard of her piloting abilities. There was a shortage of qualified male pilots and Colonel Tunner realized that women could help fill this gap. In June 1942, General Harold George approved the use of women pilots to ferry airplanes.

At first, the women were to be a part of the Women's Army Corps (WAC). However, because there was no WAC provision for flying officers or flight pay, another organization was formed — the Women's Auxiliary Ferrying Squadron (WAFS) — and the women pilots were hired as civilians.

The WAFS recruits had more stringent qualifications than their male counterparts, they were hired at reduced pay, and allowed to ferry light planes only. In the earliest stages of women's acceptance to replace the men ferrying pilots, requirements for women to belong to Nancy Love's original WAFS program were a commercial pilot's license and 500 hours' flying time.

Nancy Love flew every airplane first. She proved women could fly the airplane, then the other women eventually got transition into pursuits. Nancy was the "guinea pig" and if she could fly it, then the other WAFS got their chance.

The WAFS was an experimental group of highly qualified, elite, civilian women pilots in the Air Transport Command (ATC), who were activated in New Castle Army Air Base, Delaware, in September, 1942. They were under the command of Nancy Harkness Love. To the initial request for women pilots, Nancy Love and the ATC received responses from twenty-three women. By October 16, 1942, those accepted numbered thirteen and ultimately reached a total of twenty-eight. The women would be used to move military aircraft within the U.S. from base to base, factory to base, or perform any other aircraft movement deemed appropriate.

On August 5, 1943, the WAFS and the women pilot training program, which had been implemented by Jacqueline Cochran, were merged into one organization called WASP (Women Airforce Service Pilots). Jacqueline Cochran was named Director of Women Pilots. Nancy Love was named WASP Executive of the Ferrying Division of ATC.

Nancy Harkness Love, WAFS founder. (Sylvia Dahmes collection)

WAFS who flew Pursuits for ATC Ferrying Divisions 1942-44

Name	Ferry Group Base	Pursuit Class/Statistics
Batson, Nancy	2	44-3 February DPZ
Bernheim, Kathryn "Sis"	2	44-13A July 15 B'ville
Bohn, Delphine	2, 5	transition at bases
Clark, Helen Mary	2	transition at bases
Donahue, Barbara	2, 3	44-4 Feb. 21 DPZ
Erickson, Barbara	2, 6	DPZ prior to pursuit school (Nov. 43)
Gillies, Betty	2	transition at bases, factory
James, Teresa	2	transition at bases, factory
Love, Nancy	2, HQ	transition at bases, factory
McGilvery, Helen	2	44-3 February DPZ
Meserve, Gertrude	2	44-2 January DPZ
Miller, Florene	2, 5, 6	44-2 January DPZ
Richards, Helen	2, 5, 6	44-2 January DPZ
Scharr, Adela	2, 3, 6	transition at bases
Scott, Dorothy	2, 5	44-2, killed at DPZ in BC-1
Sharp, Evelyn	2, 6	transition at bases, killed in P-38
Towne, Barbara	2, 6	DPZ prior to pursuit school (Nov. 43)

Legend DPZ = Palm Springs Pursuit School; B'ville = Brownsville Pursuit School

WAFS Ferrying Base Squadron Leaders Barbara Erickson, Nancy Love, Delphine Bohn, Esther Manning and Betty Gillies look over collection of WASP photos and documents at a display in Ohio. (WASP Historian Collection)

WAFS Squadron Leaders Nancy Love, Nancy Batson, Betty Gillies and Adela Scharr are the first to attend Officer Training School. (Not shown is B. J. Erickson)

Gertrude "Gert" Meserve wearing a flight suit. Note the pocket on the right knee for carrying maps and log books. (Gertrude Meserve collection)

WAFS Barbara Erickson poses with a P-38. (Barbara Erickson collection)

Early June 1943 baymates at Sweetwater. Fran Snyder, Ann Waldner, Lana Cusack, Franciszka Radziwar-trainee, May Timothy-trainee, Eleanor Feeley, Betty Scantland. Three went on the Air Transport Command and successfully completed Pursuit School.

WASP practice being officers after O.T.S.

Bell Kingcobras for Red Air Force, to be ferried to U.S.S.R. through Alaska. (Jary Johnson collection)

Jacqueline Cochran, Director, Women's Flying Training. ("The Logbook," yearbook)

Women Airforce Service Pilots

As the Women Airforce Service Pilots (WASP) organization came into being, the Women's Flying Training Detachment under Jacqueline Cochran began in November 1942. In 1943, Jacqueline became the Director of the WASP training program.

Jacqueline Cochran

Jacqueline Cochran has been called the "First Lady of Flight," possibly because she achieved more "firsts" in aviation history than any other person, man or woman. She holds more international speed, distance and altitude records than any other pilot. The highlights of her career take up three pages of the autobiography written with Maryann Bucknum Brinley, published in 1987 after Jacqueline died in 1980.

Jacqueline championed the cause of women pilots against the mores of the times and eventually, not without many setbacks, was able to take North American women pilots to fly for the British. She then returned to establish the school to train women pilots to fly for our own United States government service.

In *The Fifinella Gazette*, March 1, 1943, Jacqueline Cochran stated: "At the start of the war, I became convinced that there was a sound, beneficial place for women in the air; not to compete with or displace the men pilots, but to supplement them. I never let up trying to establish, in practice, the birth of my idea. I flew a bomber to England, partly to bring out the point, and partly to see what the English women pilots were accomplishing and how they were organized. On my return, I worked with the General Staff of the Ferrying Command on a plan which later developed. The time, just then, did not seem opportune, so with the blessings of our own authorities, I took 25 women pilots to England where they have been doing a fine job flying operational equipment behind the lines, including Hurricanes, Spitfires, and two-motored Bombers. Now we are on the verge of seeing this whole dream blossom into reality in a truly big way. The Women's Flying Training Program has already approached the proportions of our entire air program prior to the start of the war. What will be the ultimate result, good and bad, will be up to the girls themselves. You of the first classes will have the real responsibility. By your actions and results, the future actions will be set. You have my reputation in your hands. Also you have my faith. I have no fear. I know you can do the job. After graduation I will be following you with proud and anxious eyes and your success will be my satisfaction."

She was an outspoken woman,

Graduating class 43-W-2 marching in review after training at Houston. The school was moved to Sweetwater and graduating ceremonies were held there.

and provoked enmity among those who disagreed with her strong views. She generously gifted the silver class wings given to the graduates of her school — classes 1 through 7, totaling 444 women. She admonished the WASP to "be ladies;" and there was a story going around that she promised to take you behind the hangar for a "lickin' " if you didn't behave like one.

She believed in the WASP, and said so in her speech to the WASP at a reunion and in her final report how proud she was of her girls of the WASP training program.

If this aggressive, intense woman had not been able to hold her own against the "big boys" in government, the women pilots would never have had this opportunity, this wonderful chance to fly.

★★★

Two former WASP converse at a reunion:

"I didn't appreciate Cochran. I didn't realize that it was her idea right from the start that woman should be given transition up to their capabilities; and General Arnold backed her up. You see, I never appreciated her. I never knew that this was one of the conflicts that she had with the Ferrying Service. They had a limited concept of the things that women were going to be permitted to do. The Ferrying Service had the biggest fight with her; did you realize this?"

"Probably it was Nancy Love."

"Maybe. They never really come right out and spell it out."

First Group

In Jacqueline Cochran's first group of trainees, the requirements were 200 hours' flying time and a commercial flying license. In the second group of women trainees, the requirements were lowered to fill the class. Of the 60 who began training in that class, only 43 graduated. Women pilots were eager and grateful for the opportunity to contribute their flying skills to the war effort. During the program 18,000 women applied.

★★★

BJ: We had to have 500 hours to enter.

Marie: Cochran told me 200, and I had 350.

BJ: I had 500, and boy did I stretch! I went down to my last penny to get in those last few hours. I couldn't get into the first class of trainees as I had to give notice since I was working for Braniff Airways at the ticket counter at Love Field checking in all the Ferry Pilots with their Class 2 Priorities.

★★★

After training for five months at Hughes Airport, the first women pilot trainees graduated at nearby Ellington Field, Houston, Texas, in April 1943.

These 23 women pilot graduates from Cochran's school (first dubbed "guinea pigs") were sent to ferry planes for the Air Transport Command and assigned five and six each to bases of the ATC 2nd Ferrying Group (FG) at Wilmington, Delaware; 3rd FG at Romulus, Michigan; 5th FG at Love Field, Dallas, Texas; and 6th FG at Long Beach, California.

The more experienced pilots from the WAFS were chosen to be squadron leaders of the women pilots at the four bases to which women were assigned. Betty Gillies was assigned to head the 2nd FG; Adela Scharr, succeeded by Barbara Donahue at the 3rd FG; Florene Miller, followed by Delphine Bohn at the 5th FG; and Barbara Jane Erickson at the 6th FG.

Ferrying bases where the women were sent were established by their proximity to aircraft factories. The base choice largely determined what aircraft the ferry pilots would fly.

The greatest need in the war in 1943 was trained men pilots to fly combat, so the WASP took new training planes to schools around the United States, moved schools to more favorable flying sites, weather-wise, and moved schools away from vulnerable East and West Coast locations to inland fields.

Marjorie M. Gray, in the first group to graduate from Jacqueline Cochran's Houston school, tells of arriving in May 1943 at the 2nd Ferrying Group base:

"When we first reported for duty to Wilmington, we came in with our new class wings, our silver wings; some of us had instrument ratings, and we all had instrument training. We had night flying and many of the original WAFS had not had any of this. They didn't have any multi-engines at that time, so there was a certain amount of resentment of us, not from all the WAFS but some of them. But I will say Betty Gillies, Teresa James and our Executive Officer Helen Mary Clark were wonderful to us. There was no problem, but one WAFS I don't think was particularly happy about us. We finally melded together, but there were some who felt a little resentful. Our class came in with all this marvelous training and they thought we maybe had been a little bit spoiled, while they, the WAFS, were working so hard. And they did. They got their faces frozen off in the winter and they really struggled in the early days before the WASP came in."

Second Group

The second group of women pilots graduated from Jacqueline Cochran's school the following month, May 1943. Forty-three women completed the six-month training program and were sent to the four ferrying bases. Some women got their first choice of assignment, but these two groups all went to ferrying aircraft. In the training program, these women had flown primary trainers, basic training planes and, in the advanced phase, had both single-engine (higher horsepower) and the twin-engine Bobcat, the UC-78, airplanes.

Acceptance

Women did not immediately start delivering airplanes when they arrived at one of the four ferrying bases. Ground school was held every day for those not out flying; marching was also part of the routine, and regular trips were scheduled for rifle and pistol practice.

In the earliest states of women's acceptance on the ferrying bases, commanding officers, the men who issued trip orders,

Twin-engine Cessna Bobcat used for instrument, navigation and cross-country training.

and even the women pilots themselves did not know to what extent women were capable of this mission. Women, by and large, were just grateful for the opportunity to fly, and to contribute to the war effort.

Commanding officers and men in control of transition schools and operations doled out the flying trips to women in light aircraft, winter trips in open cockpit airplanes, and trips that had difficult transportation in returning to home base. Most women did not complain since they had joined the WASP to fly, and they were given this opportunity. Some ferrying group operations were more reluctant to give women flying deliveries than other bases.

What were men to do with seven, then five more women pilots, then seven more women civilians who arrived at an Army base to fly airplanes at this entirely total **men-pilot** ferrying base? By edict, the bases were required to accept them; but it took another directive from the top to get them into the air. General William H. Tunner to the rescue!

General Tunner championed the women pilots. As head of the Air Transport Command, he gave all bases an order that the women pilots were to be treated equally and advanced in training in a one-woman-to-ten-men ratio. The Fifth Ferrying Division at Dallas was reprimanded for being slow to upgrade women pilots into heavier aircraft.

Only time and proven ability gave women pilots the opportunity to fly every type of aircraft that United States factories turned out during World War II.

Some women had impressive backgrounds in aviation prior to World War II. Opportunities to advance to heavier aircraft were given to those women who had more experience; occasionally transition to heavier airplanes was given on the basis of seniority. In 1942, as today, women had to prove their ability to fly.

The women advanced as their record testified to successful missions, and Tunner's directives to the bases enabled women to have more opportunities and more equal treatment in advancement to heavier airplanes. Women chose, or were chosen, to fly bombers or fighters; usually not both.

Overseas Delivery

The first summer of the war, General Tunner set up the route from Great Falls, Montana, to Fairbanks, Alaska, by which five thousand planes were delivered to our then ally, Russia.

WASP wanted to take the aircraft on to Fairbanks, Alaska, but this delivery was made by the men stationed with the 7th Ferrying Group at Gore Field, Great Falls, Montana. Women pilots in the USA wanted to meet the Russian women ferry pilots the WASP had heard about who came across Siberia to pick up the airplanes the WASP had ferried from the Bell factory to Great Falls, Montana. The WASP were given various excuses for not being allowed to fly the ship on to Alaska: 1) No facilities for women on the Alaska run; 2) Too hazardous (though men did it; and Beryl Markham and Amelia Earhart, among others, had flown under similar conditions); 3) Women by congressional edict were limited to the continental USA, yet several WASP based at Romulus, Michigan delivered ships to Canada since they were close to the border.

P-39 going to Russia on lend-lease program. (Note red star on fuselage and wings.)

Though no WASP-piloted overseas flight was ever made, the women did a magnificent job at home. They freed many men for overseas deliveries. They were checked out in the most difficult of all ships, the pursuits. They brought two thousand P-47s away from the factory. No men were needed there at all. By the end of September 1944, WASP were delivering three-fifths of all pursuit planes.

Housing

Since the women pilots were civilians, paid by civil service, there was an awkward period of adjustment for at least the first three groups of women who arrived at these four ferrying bases.

The women lived wherever they could find a room off base. Gasoline rationing was in effect and transportation was difficult.

Women were eventually housed on the bases, and given opportunity to eat in the Officer's mess hall, and allowed access to the Post Exchange (PX) and the Commissary (general merchandise). Later the WACs came and were billeted with the women pilots at Dallas.

WASP WINGS

WASP, then known as WAFS, assigned to Air Transport Command, Ferrying Division, wore Air Transport Command Pilot Wings until the regulation WASP Wings were issued.

Official wings were provided for WASP graduates in December, 1943. When officials of the AAFFTD at Houston realized, shortly before the graduation of Class 43-W-1, that the graduates would receive no wings, the unofficial insignia were hastily contrived for Air Corps Pilot wings available at the Ellington Post Exchange, redesigned by Leoti Deaton and Lt. Alfred Fleishman to make them suitable. The Air Corps insignia was replaced by a smooth shield bearing the class designation with a scroll above the shield bearing the detachment designation "319". The detachment designation was changed to "318" beginning with Class 43-W-3. The graduates of classes 43-1 through 43-7 received their wings as a gift from Director Jacqueline Cochran.

Class 43-W-8 received the first official lozenge adorned wings, in a shining finish which later gave way to the satin finish lozenge.

At the Curtiss Factory, Buffalo, N.Y., WASP Betty Jane Bachman and Betty Whitlow check the logbook of the Kittyhawk P-40 behind them. Uniforms were not issued until Summer 1944; gloves were never issued. Curtiss PR photo. (Betty Whitlow collection)

Kay Gott sports an officer's olive-drab shirt and officer's pinks since WASP uniforms had not been issued yet. Harness straps over her shoulders support her parachute. Note the Air Transport Wings over her left pocket, and the Love Field, Dallas, Texas, photo identification on her belt.

Uniforms

Some hilarious situations developed as the civil service women pilots flew on aircraft deliveries without uniforms for some months. The military police (MPs) were aware of impersonators, and the women pilots escaped the "brig" at times by keeping their military orders and their ID (identification), issued by their base, with them at all times. The first women who graduated in uniform did so on February 11, 1944.

When uniforms were finally designed and fitted, it was impossible to take much clothing in pursuit airplanes since there was no room to carry luggage. The only space in the small cockpit had to hold the essentials of a pistol with a clip of shells to defend the "secret" equipment carried on the aircraft in case you were forced down, and maps to cover the entire USA, or at least from the factory to port of embarkation. Orders and papers on the aircraft had to be carried, as well as the logbook for the airplane.

Personal things were also carried: toothbrush, soap (a scarce item during the war, impossible to find in most hotels and never in barracks), hairbrush, comb and alarm clock (one could not depend on a hotel call; many hotels did not have room phones; or at a base there was no one to call the pilot).

✯✯✯

K: Here is the bag that I still have. I traveled two weeks one time. All the luggage I had was a small bag. My little brother Richard gave me a smaller case, which held a bar of soap (you know we never had soap in any of the hotels), my toothbrush, paste, and some deodorant, an alarm clock, bobby pins and hairbrush. You can live two weeks on that!

Kay Gott, Margery Taylor and Ann Johnson, 5th Ferrying Group, Love Field, Dallas, Texas, show off their new uniforms in February 1944. (Smithsonian collection)

Kay Gott poses for professional picture in new dress uniform jacket. ATC buttons were removed on each shoulder to avoid parachute straps digging them into her shoulders.

Personal Items

Since rationing was in effect, it was difficult to obtain some distinctly feminine articles. Metal was used in war, and bobby pins were counted and guarded. An iron for your clothes was precious. An alarm clock was a necessity. Kotex was available at every base since the men ferry pilots had wives, and there were civilian office girls and nurses. Remember, tampons had not yet been invented.

Cigarettes, liquor and candy were rationed. Movies were cheap, and shown on most bases, though not all, and not every day.

It was wonderful scuttlebutt to pass on such juicy items as which refueling stop had what hard-to-get item and which overnight stop had overnight laundry service. You listened to a seasoned pilot relate details of pressing your pants by putting them overnight, creased correctly, under the mattress, and how to get a handkerchief "ironed" by pressing it wet onto a mirror.

Morse Code

Many hours were spent learning the Morse Code dot (•) and dash (—), and practicing with flashlight at night on opposite ends of a long empty hanger with blinking dot (short flash) and dash (long flash). (See Appendix C.)

SOS, the universal code for help, is ••• — — — •••, just as "Mayday" is the voice call. "V" for Victory in Morse code is ••• — and was used in many patriotic ads during World War II.

A radio-transmitted "A" (•—) signal was heard in Northwest and Southeast quadrants, and the N (—•) signal was heard in Northeast and Southwest quadrants, with the station signal being given at set intervals. When flying away from the station, the signals became fainter. The loudest signals were over the station. The "cone of silence" was **at** the station.

It was imperative to know where the landing field was in relation to the radio transmitter, and this chart was published and carried by every pilot. It was no fun to be lost in the air. Any pilot who has flown any hours may say he was never "lost," but he or she might admit to being a little "confused" at times. "On the beam" was a common saying for everything is "right on."

Link/Instrument Training

Link training was given at Jacqueline Cochran's school (Women's Flying Training Detachment). Link, or instrument training, was a requirement for all pilots to keep up competence in flying without visual reference to the ground. The Link is a box with stubby wings on a rotating base. It has a complete aircraft instrument panel, and an electronically controlled ink pen marks the map at an adjacent desk to record your flight. Essentially the Link Trainer, now called a flight simulator, is a machine, and the desk operator can control the turbulence, wind direction and force, fuel gauge, and the radio signals that the pilot hears under the hood. The box goes completely over your head, shutting out all external light.

Hazel Ying Lee in a Link trainer.

After an interval, instrument training school was opened to women pilots; and training began, not in simulated Link Trainers, but in a real, twin-engine airplane, with a green curtain around the trainee and a visual reference left open for the instructor-observer.

Parachutes

"From time to time, parachute harnesses and weights have been discussed. This department finds no evidence of breast irritation when the harness has been properly fitted. Chutes have been fitted to small, medium and large. The ideal would be individual fittings with the subject sitting. Obviously, the cross piece should not be fitted below the breasts as in case of a jump there may be severe trauma sustained."

"Several women have 'hit the silk' successfully and have sustained no physical harm aside from a little generalized soreness. The weight of the Army seat pack parachute seems excessive for small girls."

Guns

M: Do you have a seat parachute on?

K: Yes. This is early at Love Field, this is the kind of identification we wore then.

M: Did you have a gun?

K: Not in that picture.

M: We didn't have to wear guns all the time. We had some planes with radar, and that's when we wore guns.

K: Well, we did when we got to flying this stuff because we had radar in it. When a plane carried "secret" equipment, a gun was issued to the pilot upon check-in and receipt of flying orders. No holsters were provided. The gun was turned in on return from that specific flight.

Relief Consideration

Women pilots had interesting situations regarding lavatory use; it was not uncommon at Wink, Texas, a refueling stop, to post a man outside the door which had no latch, or to have the men check inside first to see if the way was clear.

In flight: "The urinary relief matter needs some definite research. Some have managed successfully with the tube in the plane. Others have been clothed so it is impossible to use the tube. The question to be solved is: Is the present tube adequate? Special consideration should be given to clothing design to provide more ready access. The Aero Med Laboratory has designed a special suit that has not had a large-scale trial yet."

"Routine" Work Day

Ferrying assignments were not scheduled on a 5-day work week. There was a war on, and everything was geared to a full 7-day work week. One day blended into the next. Frequently you forgot which day it was, but never the **date**, since the Flight Logbook for the aircraft you were delivering had to be kept accurately.

Often it was 6 a.m. to wake up, eat breakfast, get to the flight line, check weather and route, get clearance to fly and send off a departure telegram back to the home base before 8 a.m., which was the standard requirement. Then it was off the ground at 8 a.m.!

And **wait** for no one. Our mission was to get the aircraft to its destination, and in **top shape**, and as quickly as possible. **There was a war on!**

Waiting

Certain characteristics are unique to ferrying: the ferry pilot must wait for a plane to fly. This is different from most other military pilots.

The ferrying pilot WASP waited for her plane. Delays occurred at the transfer points. Some flights were long; others, short. It took time to return to base, or to another ferrying assignment. General Tunner emphasized safety — delivering the airplane intact — so there were delays due to bad weather. A great deal of air time logged in ferrying depended on chance.

Transportation

If there was enough demand, such as many airplanes ready at one time to be moved from the factory, and enough ferry pilots available to transport them, rather than going by commercial airlines from bases to the factories, the military would send their transport plane (TARFU) loaded with ferrying pilots to pick up their airplanes. TARFU, later called MATS (Military Air Transport Service) often involved a very uncomfortable, long tedious trip in bucket seats, no hot coffee, no toilets (except a coffee can) and no food, no water and, of course, no cute steward(ess).

Can you see why military transport to the factory was to be avoided, if at all possible? One way to do this was to report in to operations at the delivery of one aircraft, and then ask if they had some airplane that needed to be moved.

★★★

In the two years and three months that women pilots served, there was much human travel in the United States. Transit busses and commercial airplanes were full. The Saturday Evening Post magazine came out all across America on Wednesdays; it was a marvel how the publishers managed that!

Nancy Batson, Class IV Pursuit Pilot, with P-38 she delivered Nov. 3, 1944. Note pistol required of all ferrying pilots when flying aircraft equipped with "sensitive" gear, such as radar, or special/secret equipment.

Bernice Batten, Ruth Daily, Betty Eames, Edna Collins (Flight Commander) with Kay Gott in cockpit of a PT-13 after landing.

As the women pilots proved useful in moving aircraft around and across America, there was a change in priority given to their return trip to base.

The President of the United States and emergency medicine had top priority for transport, and such was the urgent need for airplanes in 1944 that all ferrying pilots had next-to-topmost priority for travel space. On occasion, the WASP "bumped" movie stars from their assigned seats.

Pilots returned to their own station for a next assignment; or if there were airplanes to be moved, pilots took transport to the nearest base to pick up another aircraft.

In-transit Accommodations

All ferry pilots had occasionally interesting situations for overnight accommodations, since some bases had no provisions for "in transit" pilots.

During a ferry mission, stopover bases were usually somewhat removed from towns, making transportation difficult or haphazard.

Men pilots had easier access to housing on bases, although some stations had cots available in nurses quarters on bases for visiting women. Housing for women transient pilots improved; but transportation from housing to meals was awkward, since the women pilots remained civilians without rank during World War II and could not easily phone for a jeep transport as the male officers could.

Some factories reserved an entire floor of rooms in a hotel for ferry pilots, as did Beechcraft and Boeing in the Hotel Lassen in Wichita, Kansas. There was space reserved for Air Transport pilots at the Commodore Hotel in New York City as well.

In days of food rationing, it was a disaster to RON in one meatless town after another as you progressed across the United States!

Early Ferrying

To become proficient in flying an airplane across the United States, a leadership role was assigned to more experienced women pilots. The women went in a group, each flying one airplane. The leader of the group rotated, so that each pilot had a chance to navigate. Sometimes three, sometimes as many as five women in five airplanes made up a flight.

WAFS Cornelia Fort was the first woman pilot to die. She was killed in a collision while ferrying Basic Trainers when a male pilot, also ferrying, was harassing her in the air, and while flying formation accidentally hit her airplane, killing her.

After that incident, all formation flying was forbidden. Therefore in group flights, the aircraft might be strung out for miles; communication between training airplanes was nonexistent. This caused some comic situations occasionally if perhaps the lead plane took the group off course and the last airplane in a flight of five had no recourse except to follow.

Edna Collins was flight leader when she, Betty Eames, Ruth Dailey, Bernice Batten, and Kay Gott were delivering PT-13s from the Wichita, Kansas factory to a western flight training base. One night, they all went across the border at El Paso to tour Juarez, except Bernice Batten. The next morning, Edna said "Bernice, you didn't go to Juarez, so you lead the next lap. From Cochise there is a railroad line going down the Mexican border, and Tucson is our next refueling stop." So they start off, and head north!

Kay recalls she said to herself,"'Oh no, we are not going the right way!' I had the next-to-the-last airplane, and Edna had the last one. In a PT-13, you were in an open cockpit plane, with no radio; and we were strung out all over New Mexico with no communication among the five airplanes. We were about a 45-degree angle off from where we were supposed to be, and we all landed at Deming, New Mexico. The sergeant ran out — he had never seen women fly before — and informed us we flew through a bombing range (ship outlines are painted on the desert floor to train bomber crews to hit their targets) and we flew through it! We could have been bombed by the planes above us."

Kay Gott in the cockpit of a Boeing Stearman PT-17 bi-plane delivered to training schools all over the southern United States.

Insurance

The WASP were civilians serving under military regulations but receiving no GI benefits. As WASP, they couldn't get military insurance because of their civilian status, but they couldn't get civilian insurance because of the risks they took.

WASP In Review

The summer of 1944 saw a diminishing number of training-type aircraft. The training of pilots had largely been accomplished.

In August 1944, there was a great need for pilots to fly fighters to delivery ports. This is further detailed in *"History of the ATC - Women Pilots."* Women who chose not to fly pursuits would be transferred out of the Ferrying Command and placed in the Training Command.

Women not in pursuits would be transferred out of the ATC in August 1944. Women were moving training planes, thus taking men out of the orderly transition pattern from trainers to bombers, according to the 1946 Revision of Doc. 55 History.

This did not exactly take place, however. Upon examining the closing day's photographs taken on or near December 20, 1944 deactivation day at the five bases where women pilots were stationed, some women pictured did not fly pursuits. Moving some lighter type aircraft was still necessary, and the Ferrying Command objective was to move aircraft!

★★★

Kay: It said in the Historic Document 55 that at the end nobody could stay in ferrying unless they flew pursuit. You had to exodus, you had to make a choice, you had to go into pursuits or out you went. Out to fly weather, tow targets or whatever. Wherever they needed you.

Marie: You mean every WASP had to fly pursuits?

K: No, everyone at Dallas, for they made pursuit airplanes at Hensley Field, near Dallas.

M: Some women were towing gliders!

K: But they were proved not satisfactory and the conclusion was, "Women lack the strength to do this kind of work." But there were only three women used in this experiment.

M: What does it take to tow a glider?

K: I don't know. I checked out in the B-18 and it's just like an old truck. You pull and you pull and you pull on the ailerons and finally the thing starts to slowly turn. That takes more strength than any ship I flew. And I got my instrument rating in that plane.

K: Now here's another interesting dilemma. It states in Historic Document 55 some guy got three WASP to tow targets: radio-control thing, and he writes pages of complaint: "This demands more than ordinary flying ability" and in his judgment this experiment showed conclusively that women were not suited for this kind of exacting work. Cochran makes a strong point at the end of her report that the women excelled in target-towing. Here are two conflicting views. It names the man who turned in the report that the women weren't good enough. The reply to his letter states: "Your report is received; however, your conclusions are on a limited sampling of girls, and on three-days' duration, and it seems to me that your sampling is too small and the time duration should be at least 30 days before you make any judgment as to the suitability or unsuitability of women to do this kind of work." And this is not from Cochran, this is from somebody else, who received the report. So it is good that she makes this statement at the end, that women did excel. But I never saw any figures. There were women lost in towtarget. Was one woman shot down? More than one? When the guys shot the plane down instead of the target?

M: I don't know.

Evidently, those girls didn't have a choice when they were sent to towing targets.

K: I think not.

M: If they gave me a choice, no way would I tow targets! No way was I going to have those guys practice on my airplane! Shooting at me with real, live bullets! You make one wrong turn and you'd be killed.

K: There was some girl killed in training on tow-targets, but it just says in our Roster "deceased." Not much information as to why or where.

Target Tow

Dora Dougherty, WASP class of 43-W-3, graduated July 3, 1943 and was assigned to tow targets for anti-aircraft squadrons in training. She flew the SBD or A-24 Douglas Dauntless, a single-engine, fast plane most often. After that, she was a radio-control drone pilot using PQ-8 Culver planes and the UC-78, a "Classified" project at the time. She and classmate Dorothea Johnson were chosen to fly the B-29 as demonstration pilots to flight crews in transition to B-29s.

WASP Disbanded

"Operations by women pilots came in to an end in December 1944. In 27 months, the women ferry pilots had completed 12,650 ferrying movements over a distance of 9,224,000 miles - a record of useful achievement and of solid contribution to the prosecution of the war.

"In their latter months, they specialized in the delivery of a type of aircraft which did not fit into the transition program, thereby facilitating the advancement of male pilots to foreign duty, combat and non-combat. That this type was especially dangerous to fly, that many of them crashed and some were killed, did not make them hesitate.

"These WASP were engaged in an outdoor activity that demanded great skill and concentration and did not permit time for introspective dreaming or for the nursing of any latent neurotic tendencies. The type of women who engaged in ferrying were distinctly balanced on the extrovert rather than on the introvert side.

"Every big bomber and every cargo plane flown by women pilots were safely delivered. In all, they delivered 12,652 planes while flying 77 different types. Yet the most we ever had in Air Transport Command was 303 WASP; at the same time we had eight thousand male ferrying pilots.

"The two-year existence of the women pilots flying coincided with the period of the Ferrying Division's greatest growth."

These women were carefully selected, seasoned pilots. They were hired as civilians whose primary duty was to free male pilots for combat. They were not really a part of the Army. The WASP bought their own uniforms, paid most of their own ground transportation and reported for duty trim and fit in Santiago blue and wearing silver wings.

They ferried military aircraft, towed air gunnery targets, and taught Army Air Force (AAF) cadets how to fly. But they were a thorn in the side of the men fliers; they had pricked the bubble of male superiority. As one AAF officer said later: "They were too damned good for their own good."

★★★

6th Ferrying Group
Long Beach, California
December 9, 1944

General H. H. Arnold
Commanding General
Army Air Forces
Pentagon Building
Washington, D.C.

Dear Sir:

For over two years the subject of the Women's Air Force Service Pilots has been one of bitter controversy, criticism, and debate. The public has heard about the WASP from the press, radio commentators, Congress, and Miss Cochran. Each of these sources has confused, distorted, and misrepresented the facts. Because you are our Commanding General and because we feel we are deeply indebted to you for the privilege of serving as members of the Army Air Force, we would like you to hear from the WASPs who, till now, have remained silent. When a person is deprived of the only weapon he possesses to combat opposition, the opportunity to prove his worth by his work, he is tempted to take up the pen. We do this with humble gratefulness that we live in a country where a WASP can write to a General.

The announcement that after December 20th we would no longer be members of the Army Air Force has left us feeling disappointed, bewildered, disillusioned and betrayed.

Very recently we were told that 2,000 more women were needed to be trained as pilots. Yet, a few months later, we were informed that the Army no longer needs any women flyers, not even those now highly qualified. We are accused of taking jobs away from thousands of male pilots. We believe the primary consideration in this respect should not be whether a pilot is male or female, but rather what is most expedient for the war effort. That girls with 35 hours should be trained, when scores of men pilots are available, seemed to us just as incongruous as the dismissal of those WASPS now capable of handling class 3P, 4 and 5 aircraft to hire men with only trainer experience. We asked only to be judged not as a group or as women, but as individual pilots on our record and ability. If there is now a surplus of airmen, we believe selection as to who shall serve in the AAF should not be on a basis of Army Pilot, Service Pilot, Civilian Instructor, or WASP but on individual merit. Since the bearing of arms in actual battle is still solely a male prerogative, we feel that whether a woman relieves a man from a cockpit or a typewriter is inconsequential as long as she handles it with equal skill.

- 1 -

Page 2
General H. H. Arnold

We are told that recent battlefront successes make our services no longer valuable. Yet, we in the Air Transport Command have been busier the past few months than ever before. Thirty-four percent of all fighter-type aircraft in the United States are being ferried by 140 WASPs. Members of the WASP Squadron here at the 6th Ferrying Group are qualified to fly twenty-one different types of planes. All members are class 3P with instrument ratings; over half are class 4P or higher.

We believe that the main issue precipitating the inactivation of the WASP was that of militarization. Ninety percent of the WASPs did not support Miss Cochran's alternative that either we be given full military status or be disbanded. We considered our Military status of secondary importance, subordinate to the war effort. We believed that as long as a combat pilot received quickly and in good condition a plane we ferried to an embarkation point it mattered not whether we did so as civilians or Officers. If the Army had considered that in the interest of efficiency, organization, and control, it was better that we be militarized, almost all of us would have accepted commissions. However, we were not in sympathy with establishing a Women's Air Force staffed by a feminine echelon of command with those at the top exploiting the organization for personal power. While we felt we were entitled to Military benefits, especially those of hospitalization and insurance, we accepted willingly our status, cognizant of the risk involved. That an ultimatum be issued in time of war that the WASPs receive full Military benefits and status or be disbanded seemed to us to have all the treachery of a Labor Union War-plant strike.

We read the account of columnists and the accusation of Congressmen which attacked our safety record, ridiculed our ability, (based on the conclusion that we did not all fly B-17's and P-38's), and labeled the proposed militarization bill "Society Legislation." We wondered if our accusers know what it is to fly a pursuit ship all day with the temperature 110 degrees inside the cockpit; deliver an open trainer in the middle of winter; ferry all day, then ride the cold bucket-seats of a cargo plane back all night; or to pack the belongings of a room-mate who flew out one day never to return. Because of these people and their publicity, we will go down in history as a dismal failure and as Time Magazine puts it "getting what we asked for." Against these people we need no defense. The satisfaction of our work accomplished is sufficient. To you who had faith in us, we feel we must justify ourselves. We realize there were undesirables in our ranks who brought discredit to our organization. We were selected with quantity not quality as the goal.

Page 3
General H. H. Arnold

    We are aware political entanglements influencing the action to inactivate the WASP was beyond your jurisdiction. We want you to know we were willing and proud to serve as pilots in the Army Air Force under any status anywhere and in any capacity. We only regret we must be kicked off the team and forced to turn in our suits in the fourth quarter of the game.

                                      Sincerely,

                                      Virginia Hill
                                      WASP
                                      6th Ferry Command
                                      Air Transport Div.

New York Herald Tribune

Gill Robb Wilson

**HEADQUARTERS
WOMEN'S AIR FORCE SERVICE PILOTS
552ND AAF BASE UNIT
(2ND FERRYING GROUP)
FERRYING DIVISION - ATC
NEW CASTLE ARMY AIR BASE
WILMINGTON 99, DELAWARE**

In reply refer to:

WASPS TO BE DEMOBILIZED DEC. 20

Of Nine Hundred of Them Two Hundred Are Experts, Needed in War Work

Demobilization of the Wasps (Women's Airforce Service Pilots) on Dec. 20 will prove costly to the war as well as to the taxpayer unless supplemental action is taken. For among the total Wasp personnel there are approximately two hundred women pilots of outstanding ability and specialized experience, irreplaceable in their present assignments without serious loss of time and money.

Provision is now needed empowering the air forces to hire in a civilian capacity any woman currently qualified to fill a specialized job for which, without additional training, no male pilot is available. Part of the unused appropriation for the Wasp training program, now abandoned, could meet the comparatively small pay roll of such an arrangement.

It well may be that some solution of this kind is already is the mind of General Arnold, air forces commander, who can always be counted on to make the most effective use of pilots, whether they be men or women.

I take up no cudgels for women pilots because they are women nor for men because they are men, but only to get on with the winning of a war, the toughness of which is not yet generally comprehended. Approximately two hundred of the nine hundred Wasps can fly expertly any plane that is used in battle today, including jet-propelled aircraft. Training of men to take their places will require a million dollars, from four to six months' time, and even then will not replace the broad experience which the women have built up on pursuit-type aircraft. I just do not think the United States is rich enough to throw away two hundred expert pilots, regardless of who they are.

For example, from one air base alone some thirty-five women pilots have ferried in less than a year more than 3,000 aircraft, of which more than two-thirds were pursuit types. From one great pursuit aircraft factory women pilots have ferried all ships built there during the last half year. From another they are ferrying more than 80 % of all ships produced. In fact, approximately 50 per cent of all pursuit ferrying in the United States is done by WASPS.

Winter is upon us. Weather for the next four months will require at least double the number of ferry pilots ordinarily needed. Aircraft production of pursuit types is at a high level and must remain so. Yet aircraft are piling up at factories for lack of ferry pilots. Freezing men in pursuit ferry specialization blocks them from further upgrading, and therefore from advanced utility overseas. What a time for the air forces to be compelled to fire two hundred top hands!

Letter written to WASP from columnist Gill Robb Wilson, published in New York Herald Tribune.

New York Herald Tribune

HEADQUARTERS
WOMEN'S AIR FORCE SERVICE PILOTS
552ND AAF BASE UNIT
(2ND FERRYING GROUP)
FERRYING DIVISION - ATC
NEW CASTLE ARMY AIR BASE
WILMINGTON 99, DELAWARE

In reply refer to:

Considering all elements involved, the demobilization of the Wasps as such is undoubtedly is indicated, particularly in view of the numerically increased male personnel available for most phases of air forces activity. However, there is no subsistute for experience. And for eighteen months a sizable number of Wasps have specialized on pursuit ferrying.

In their two years' existence the Wasp organization as a whole has logged better than half a million air hours on almost every type of mission exept combat. In ferrying they have flown all aircraft from the biggest bomber to the hottest pursuit, and their low accident record is a joy to behold. But, as previously indicated, the question is not of demobilizing the Wasps, but of firing the experts among them who can not be replaced in full and whose partial replacement will be expensive and slow.

Activities of women pilots in war service began in September, 1942, when a group of women with extensive civil flying experience was employed by the Ferry Command which later merged with the Air Transport Command.

In the autumn of that year, with high combat casualties possible, the air staff was compelled by logic to inaugurate a womens' training program. Some one had to find out whether women could adapt themselves to military flight training in sizable numbers if the country became hard pressed for aviators. In those days the Luftwaffe was riding high, wide and handsome and none could know what would be required to lick the Nazi and the Jap. Had General Arnold not initiated a women's pilot training program for potential release of men for combat duty, he would have been derelict, and he has never been that.

While the women's training program was in progress the men's training program was likewise in full swing. Gradually it was seen that large numbers of women pilots would be unnecessary, for Air Force casualties were lower than expected. Eventually the decision to abolish the women's program was reached. It is not the prerogative of the writer to question that decision, for men are available to do most of flying chores which the general run of the women pilots would have done.

But men are not available without a lot of additional training to do some of the specialized flying for which some two hundred women are not only prepared but at which they have accumulated experience that is more than satisfactory.

The women's training program proved that the female is no deader than the male when either crack-up and that the women do constitute a reserve air power personnel if we ever need them again.

Whatever else may be said about the Wasp organization, the first thing that be said is that they did a solid job and sometimes had to take low blows without squealing. Any girl who got through the program can really fly an airplane. And those who have been pushing "pea-shooters" about the land so long and so safely are stars in any sky.

If we still do not need them a lot of top-side people in the service whose judgment I respect are 100 per cent off the beam. Great Britian still need her top-flight women pilots. So does Russia. And--unless I'm just a chronic pessimist about this war---so does the United States.

APPENDIX D. ADDRESS BY
GEN. HENRY H. ARNOLD, COMMANDING GENERAL
ARMY AIR FORCES
TO
LAST W.A.S.P. GRADUATE CLASS, 7 DECEMBER 1944

I am glad to be here today for a talk with you girls who have been making aviation history. You and all WASP have been pioneers in a new field of war-time service, and I sincerely appreciate the splendid job you have done for the AAF.

You, and more than nine hundred of your sisters, have shown that you can fly wingtip to wingtip with your brothers. If ever there was a doubt in anyone's mind that women can become skillful pilots, the WASP have dispelled that doubt.

The possibility of using women to pilot military aircraft was first considered in the summer of 1941. We anticipated then that global war would require all our qualified men and many of our women. We did not know how many of our young men could qualify to pilot the thousands of aircraft which American industry could produce. There was also the problem of finding sufficient highly-capable young men to satisfy the demands of the Navy, the Ground Forces, the Service Forces, and the Merchant Marines. England and Russia had been forced to use women to fly trainers and combat-type aircraft. Russian women were being used in combat.

In that emergency I called in Jacqueline Cochran, who had herself flown almost everything with wings and several times had won air races from men who now are general officers of the Air Forces. I asked her to draw a plan for the training and the use of American women pilots. She presented such a plan in late 1941 and it formed the basis for the Air Force's use of WASP.

Frankly, I didn't know in 1941 whether a slip of a young girl could fight the controls of a B-17 in the heavy weather they would naturally encounter in operational flying. Those of us who had been flying for twenty or thirty years knew that flying an airplane was something you do not learn overnight.

But, Miss Cochran said that carefully-selected young women could be trained to fly our combat-type planes. So, it was only right that we take advantage of every skill which we, as a nation, possessed.

My objectives in forming the WASP were, as you know, three:

1. To see if women could serve as military pilots, and, if so, to form the nucleus of an organization which could be rapidly expanded.

2. To release male pilots for combat.

3. To decrease the Air Force's total demands for the cream of the manpower pool.

Well, now in 1944, more than two years since WASP first started flying with the Air Forces, we can come to only one conclusion--the entire operation has been a success. It is on the record that women can fly as well as men. In training, in safety, in operations, your showing is comparable to the overall record of the AAF flying within the continental United States. That was what you were called upon to do--continental flying. If the need had developed for women to fly our aircraft overseas, I feel certain that the WASP would have performed that job equally well.

Certainly we haven't been able to build an airplane you can't handle. From AT-6's to B-29's, you have flown them around like veterans. One of the WASP has even test-flown our new jet plane.

You have worked hard at your jobs. Commendations from the generals to whose commands you have been assigned are constantly coming across my desk. These commendations record how you have buckled down to the monotonous, the routine jobs which are not desired by our hot-shot young men headed toward combat or just back from an overseas tour. In some of your jobs I think they like you better than men

I want to stress how valuable I believe this whole WASP program has been for the country. If another national emergency arises--let us hope it does not, but let us this time face the possibility--if it does, we will not again look upon a women's flying organization as experiment. We will know that they can handle our fastest fighters, our heaviest bombers; we will know that they are capable of ferrying, target towing, flying training, test flying, and the countless other activities which you have proved you can do.

This is valuable knowledge for the air age into which we are now entering.

But please understand that I do not look upon the WASP and the job they have done in this war as a project or an experiment. A pioneering venture, yes. Solely an experiment, no. The WASP are an accomplishment.

We are winning this war--we still have a long way to go--but we are winning it. Every WASP who has contributed to the training and operation of the Air Forces has filled a vital and necessary place in the jigsaw pattern of victory. Some of you are discouraged sometimes, all of us are, but be assured you have filled a necessary place in the overall picture of the Air Force.

The WASP have completed their mission. Their job has been successful. But as is usual in war, the cost has been heavy. Thirty-~~seven~~ eight WASP have died while helping their country move toward the moment of final victory. The Air Forces will long remember their service and their final sacrifice.

So, on this last graduation day, I salute you and all WASP. We of the AAF are proud of you; we will never forget our debt to you

HEADQUARTERS OF THE ARMY AIR FORCES
Washington 25, D. C.

22 February 1945

TO ALL FORMER WASP:

    This Headquarters has recently been advised that a circular, relating to the possibility of entering the Women's Army Corps as rated flying officers, has been directed to a number of former members of the WASP.

    While the sponsorship of this circular is not known, it should be pointed out that it was prepared and distributed without the knowledge of this Headquarters and without coordination with the Women's Army Corps. Undoubtedly many of its recipients are hopeful that, through the WAC, they may be able again to contribute to the war effort in a flying capacity.

    The Army Air Forces investigated every possible avenue for utilization of its women pilots before inactivating the WASP on 20 December 1944. The situation which then existed has not changed. There are sufficient trained male pilots to handle all the domestic flying assignments of the AAF. There is not now any military requirement for women pilots, nor is there any foreseeable requirement for them in a pilot's capacity.

    That lack of requirement provides an answer to the question of the possible use of former WASPs as pilot officers with the Women's Army Corps. There are now no direct commissions in the WAC, officer status being given only to a very limited number of enlisted WAC who are selected from the ranks and then graduate from an officers' candidate school.

    No hope can be held out to former WASP of a return to military flying duties through the WAC or through any other civilian or militarized arrangement now visible.

    The WAC does have a need for volunteers, who do good service in the war effort in various capacities which, however, do not include piloting.

    It seems only fair that all former WASP understand the exact situation. Many have delayed obtaining other war service employment in the hope of recall to flying duty with the AAF. Unfortunately, that hope has no foreseeable basis in fact.

    The patriotic motive which has guided former members of the WASP to seek to reenter the service of their country in the capacity for which they have been trained is deeply appreciated.

Sincerely yours,

Jacqueline Cochran
JACQUELINE COCHRAN

Commendation

Women's Air Force Service Pilots of the WASP Squadron of this Command are hereby commended for extraordinary achievement accomplished during 22 months service between 16 February 1943 and 20 December 1944. Expanding from a nucleus of five pilots to an efficient organization of 70 fully qualified ferry pilots, they flew 21 types of military aircraft on regularly scheduled domestic ferrying delivery flights to all parts of the nation. In the month of January, 1944, 65 WASPs delivered 215 aircraft from these runways and logged 2,031 hours first pilot time. In June, 1944, 35 WASPs delivered 78 aircraft, 67 of which were of the high speed single engine pursuit type. During the nine month period between January and November, 1944, WASPs of this command delivered a total of 980 military aircraft, an accomplishment which not only reflects credit upon themselves but upon the Sixth Group and the Ferrying Division of the Air Transport Command through exceptional flying skill, stamina and devotion to duty. WASPs of this command won a place of equality with male pilots and performed their assignments without special privilege or dispensation due their sex. Performing a man's job with courage, diligence and ability, they contributed materially to the large scale delivery of the nations military aircraft to the battlefronts. And this contribution has won them a high place in the annals of women in the war - - - - -

Earl D. Johnson
Lt. Col. A.C.
COMMANDING

Commendation to WASP of 6th FGp, Long Beach, CA (Virginia Hill collection)

Women in Pursuit

What was so different about this unique job of ferrying fighters, rather than ferrying training airplanes that most women pilots had done for nine months to a year previously?

All in all, women pilots were engaged for over a full year ferrying fighters, at the time labeled "pursuits."

Qualities of Women in Pursuit
They were good-looking, interesting people, highly competent, and held themselves in a confident manner. They were assertive in that they were able to hold their own in a profession that was largley ruled by men. They were individualistic — no two alike. They were loners in that they were able to function successfully alone, on their own resources. Mostly, they were fun-loving and they all proved dependable. They got the job done.

They were all proud of their flying ability, but tended to be overly modest in mixed company, and very casual about their accomplishments. One had to pry out of them what exactly they did. After all, they were "just moving airplanes" while men were facing, or about to face, the enemy, perhaps to be shot down and possibly killed in combat in these planes.

The women who flew pursuits were physically fit, and had the endurance and stamina for the frequent refueling stops; landings were demanding. Women pursuit pilots succeeded in having the smooth coordination and a "caressing" touch on the controls.

Nearly every woman had a year to nine months ferrying before going to pursuit flying.

Some of the pursuit-qualified women pilots were called "Eager Beavers." The eager pilots had one thing in common — they loved to fly!

These women enjoyed life and laughter, and felt good about what they did. They felt as though they made a valuable contribution with their flying skills. Each job they did freed another man.

The women pilots who came through deaths of their comrades felt that accidents happened to someone else—never to them. The ones who lasted and came through disasters, and ones who lived through the fear of death, had a high respect for danger. Did they live with fear? No. Did they ever fear? Of course, all pilots do. A good pilot is one who responds correctly to overcome the cause of the fear, be it engines that quit, tires that blow, unexpected weather — whatever the cause. These pilots were thorough, skilled in their flying, calm in emergencies: did the right thing and responded in time — almost automatically — to avert tragedy.

Palm Springs Pursuit School

Barbara Erickson and Barbara Towne were the first women pilots to be sent from the 6th Ferrying Group to Palm Springs before the school officially opened to train the ferry pilots in the pursuit planes in December 1943.

There were eight classes with women at Palm Springs, though the records are nonexistent and sketchy as to who was in the classes, dates of graduation, and photos missing or perhaps never taken.

Mary Lou Colbert was in the last class — the lone woman at Palm Springs, California. The school was moved to Brownsville, Texas, in April 1944; and Palm Springs became the 21st Ferrying Division. Evelyn Sharp, 6th FG was to have been the first Women Squadron Leader at Palm Springs, but she was killed in a P-38 at New Cumberland, Pennsylvania on April 3, 1944 before she assumed command. Byrd Granger then became the leader, followed by Mary Lou Colbert.

Weather was so foggy at Los Angeles and Long Beach, the short hop to Palm Springs was useful to get the planes out of the coast when the weather did break. All 12 of the women at Palm Springs after it became a ferrying base did fly the pursuits — the only WASP base where all women flew pursuits.

Brownsville Pursuit School

One statistic is that there were 96 graduate WASP from Brownsville. The Brownsville Pursuit WASP have been easier to find than the women who graduated from Palm Springs. Not without difficulty, however. As the photographs of pursuit classes came, it was gradually realized that these women pictured did not all graduate, for some reason or another. Since the earlier photographs were all graduates, it was assumed that the photos would all be the graduates. Corresponding with many, many WASP it was learned, from orders, and from the women themselves, that many of the pictures in this book show women who arrived at the school, not necessarily just those who successfully completed the tough course. This documentary was only begun in 1987, and many of these women are now dead. After all, this remarkable event in our lives did take place in 1942, now some 51 years ago!

The school became established and women were sent to learn to fly in four fighter aircraft: P-40s, P-47s, P-51s and P-39s. Some mystery shrouds the why and how women were chosen for this school. Presumably the women with the most air experience were sent first; but as their orders show, the women did not always get to be the pursuit class attendee. Perhaps the explanation might be that the woman pilot on orders to attend pursuit school was away on a trip delivering another airplane; and since her place was vacant, it was taken by an available woman pilot from another base. Later, there was some attempt to send all women pilots from each base to the school.

Evelyn Sharp was assigned as first Women's Squadron Leader at Palm Springs; however, she died in a P-38 crash prior to assuming command. (Barbara Erickson collection)

HEADQUARTERS
568TH AAF BASE UNIT
(4TH OPERATIONAL TRAINING UNIT)
FERRYING DIVISION - AIR TRANSPORT COMMAND
BROWNSVILLE MUNICIPAL AIRPORT
BROWNSVILLE, TEXAS

OPR/GTR/JFJ/bl
15 April 1944

SUBJECT: Schedule for Class 44-9.

TO : All Concerned.

   1. The following is the schedule for Pursuit Pre-Flight Training beginning Monday, 17 April 1944, for Flight A and B, Section I.

Monday, 4-17-44   <u>Flight B</u>
Time	Subject	Instructor	Room
0730 - 0820	Orientation	Capt Jennings / Capt Rigney	T-53A
0830 - 0920	Lecture by Safety Officer	Capt Broome	T-53A
0930 - 1020	SCR - 522	Sgt Marietta	T-55B

<u>Flight A</u>
Time	Subject	Instructor	Room
1430 - 1520	Orientation	Capt Jennings / Capt Rigney	T-53A
1530 - 1620 ✓	Lecture by Safety Officer	Capt Broome	T-53A
1630 - 1720	SCR - 522	Sgt Marietta	T-55B

Tuesday, 4-18-44   <u>Flight B</u>
Time	Subject	Instructor	Room
0730 - 0820	Cockpit Familiarization P-47	Lt Pollard	T-53A
0830 - 0920	Pre-Flight P-47	Lt Pollard	On the Line
	Pre-Flight P-39	Lt Roth	Line
0930 - 1020	Cockpit Familiarization P-39	Lt Roth	T-54B

<u>Flight A</u>
Time	Subject	Instructor	Room
1430 - 1520	Cockpit Familiarization P-47	Lt Link	T-53A
1530 - 1620 ✓	Pre-Flight P-47	Lt Link	On the Line
	Pre-Flight P-39	Lt Baughman	Line
1630 - 1720	Cockpit Familiarization P-39	Lt Baughman	T-54B

Wednesday, 4-19-44   <u>Flight B</u>
Time	Subject	Instructor	Room
0730 - 0820	Cockpit Familiarization P-40	Lt Calkins	T-53B
0830 - 0920	Pre-Flight P-40	Lt Calkins	On the Line
	Pre-Flight P-51	Lt Deckner	Line
0930 - 1020	Cockpit Familiarization P-51	Lt Deckner	T-54A

<u>Flight A</u>
Time	Subject	Instructor	Room
1430 - 1520	Cockpit Familiarization P-40	Lt Paylan	T-53B
1530 - 1620	Pre-Flight P-40	Lt Paylan	On the Line
	Pre-Flight P-51	Lt Miller	Line
1630 - 1720 ✓	Cockpit Familiarization P-51	Lt Miller	T-54A

                    P-47
Thursday, 4-20-44   <u>Flight B</u>
Time	Subject	Instructor	Room
0730 - 0820	Precautions & Emergencies	Lt Pollard	T-53A
0830 - 0920	Systems	Lt Pollard	T-53A
0930 - 1020	Pratt & Whitney Engines	Mr Mott	T-53A

-1-

Typical Training Schedule, first page only shown. (Betty Shea collection)

Lana Cusack prepares to fly an AT-6 used in liaison efforts. (Lana Cusack collection)

On examining the school photographs, comparing the class as of graduation, and the bases from which each girl was sent, the selection of women pilots seems haphazard. At some bases, women pilots were asked if they would like to ferry pursuit aircraft; at other bases, they were just sent because there was a request for a quota of women. The memo from General Tunner stated that women were to be given opportunity for advancement on a ratio of 1 woman to 10 men. The statistics and the photographic evidence do not bear this out, however.

Time at school was lengthened in later classes since the women and men did not have as much flying background as earlier classes.

The washout rate for women in pursuit school was high. Some women appear in two different classes. Was this due to illness? Or perhaps she did not have enough flying experience and wanted to try again at a later date? Some women got to chose, and other bases just sent women, whether they wanted to go or not. Some women appeared on orders to attend, but did not go — possibly they were away from base at the time the school started as were Alice Lovejoy, Lillian Conner, Marjorie Gray, Virginia Sweet, Jill McCormick. Some women are pictured in two different classes, and some women who attended were never photographed. This made the research extremely difficult, for the searcher believed that the photos were graduation pictures, and they proved not so.

To add to the confusion, some women learned to fly pursuits at the factories and never attended pursuit school. The best records came from the women themselves; and since this search was only begun 43 years later, some of the WASP involved had died, and records were obtained from their flying WASP friends, from Operations Orders, or from their Squadron Leader.

Pursuit school was closed to women in October 1944. Only a few men-only classes were held after that date.

Base Assignments
Men

After pursuit training, the men were transitioned to four-engine school. The men rotated in to co-pilot four-engine planes overseas. From Dallas, Texas, the men ferried B-17s and B-24s to Caracas, Venezuela; on to Belem, Brazil; refueled at Ascension Island; then to Dakar, Africa and north to England. They were away from the base for long periods of time on many flights.

Women

Each woman who went on to fly pursuits served eight to ten months to a year flying lighter aircraft, liaison aircraft and trainers. Some women were upgraded to heavier aircraft earlier than others. Some women were given a choice at some of the bases whether to fly multi-engine or to go to pursuit school. Some women were **not** given this choice.

Women in pursuit ferrying were the only steady, ferrying corps the Commanding Officers of the bases had permanently assigned to the four, then five bases. Women in pursuit were stationed under the Air Transport Command to these five ferrying bases by 1944:

2nd - Wilmington, Delaware
3rd - Romulus, Michigan
5th - Dallas, Texas (Love Field)

6th - Long Beach, California
21st - Palm Springs, California

There were no women pilots stationed at the 7th Ferrying Group, Great Falls, Montana, or 4th Ferrying Group in Memphis, Tennessee.

The Republic Aviation Corporation at Farmingdale, Long Island, New York and Evansville, Indiana factory had a contingent of women pilots who rotated in on temporary duty assignments (TDY).

Transitions

Some WASP did not attend pursuit school, but transitioned at their base, or learned to fly pursuits at the factories.

The Republic factory gave experienced and qualified women transition on the P-47 at the factory. The Niagara Falls Bell aircraft factory gave transition checkout in the P-63 Kingcobra to those women and men pilots who had previous flying time in the earlier model Bell P-39 Aircobra.

The Lockheed aircraft factory was at Long Beach, California, and, in the early days, some of the women went on TDY just to be checked out in a P-38. Helen Mary Clark went from Wilmington, Delaware for this purpose. Later, at the Lockheed modification plant in Dallas, Texas, women were given transition along with the men in ground school, cockpit checks and practice. There was a great need to move these airplanes to ports of embarkation, and women helped accomplish this.

Base Rules

Rules were different for pursuit pilots at each base where the women worked. At Long Beach 6th Ferry Group you could fly fighters at night if you were checked out.

At Dallas 5th Ferry Group, night and instrument flying in fighters was forbidden; instrument training was kept up, but used only in emergencies in fighters. Night

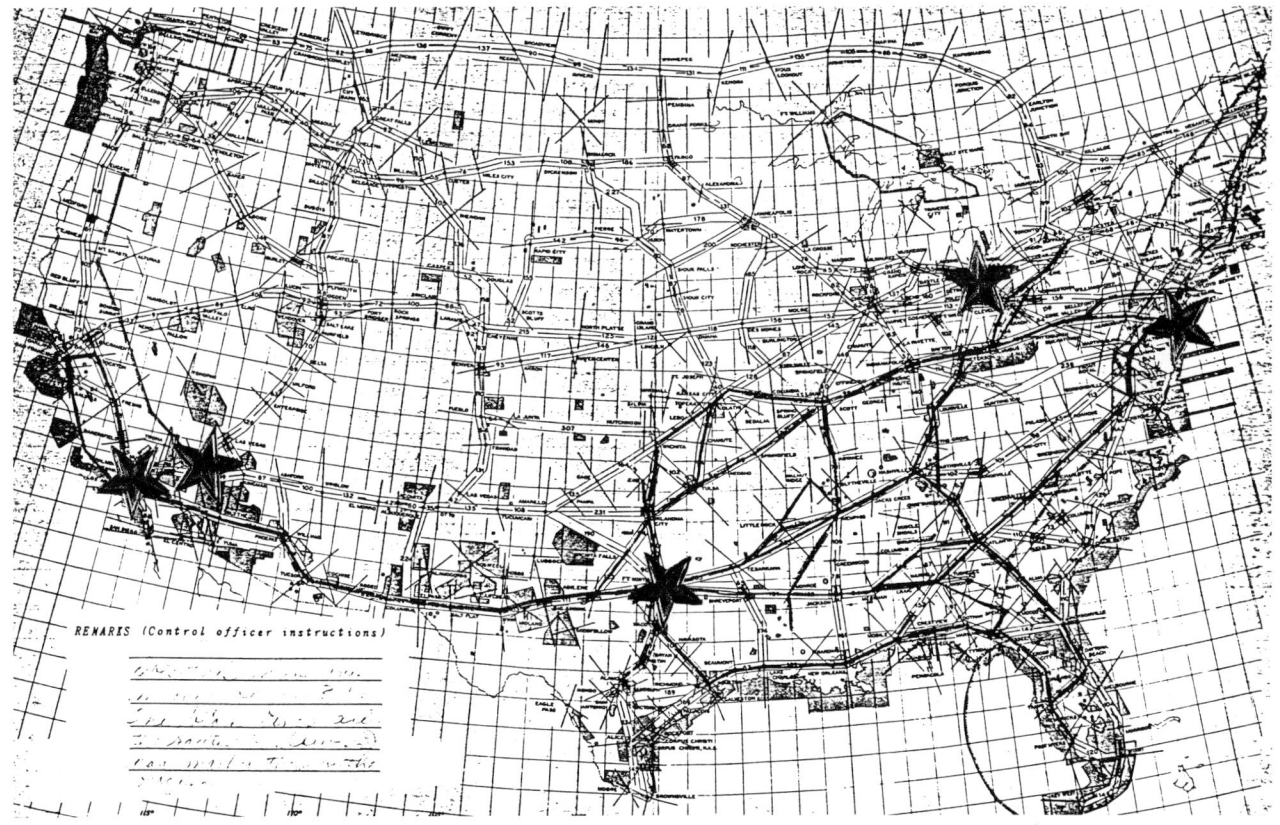

WASP Air Transport Command bases (where the stars are) and the air routes they flew.

check-outs were not even given at Dallas. 5th Ferrying Group fighter deliveries were made alone, even over unfamiliar routes.

The rule at Dallas was: "Get the plane to its destination, someone there needs it; don't wait for anyone; and no flying together!"

Dallas had a regulation that the pursuit pilot must land 30 minutes before sunset, with half-an-hour of fuel reserve, and be off at 8 a.m. the following morning or else state in your daily telegram the reason why. The usual reason for not flying was one word: weather.

At Dallas, the shuttle from the nearby North American Aviation P-51 Mustang factory at Hensley Field to Love Field at Dallas was a boring, time-consuming task with little air time and long waits between those short flights.

Betty Shea reports that at Romulus, 3rd Ferrying Group it was customary to follow a Flight Leader as she did from Niagara Falls, New York, to Great Falls, Montana, her first trip on that route across the mountains and plains of America, delivering Bell P-39s and then P-63s.

Qualification Ratings

In the Army Air Corps during the years the women flew, there was a Qualification Record carried by each pilot, which attested the type of aircraft that pilot was proficient to fly. The classification was by

Class	Type of Aircraft
I	PT, BT, AT-6, BC-1, C-43, C-61, C-64, C-71, C-72, C-81
II	AT-7, AT-8, AT-9, AT-17 (C-78), AT-10, AT-11, AT-15
III	A-29, OA-9, OA-10, B-23, B-18, C-33, C-39, C-47, C-53, C-56, C-59, C-60, C-57
IV	B-25, B-26, C-46, A-20, A-30, P-38, P-61
V	B-17, B-24, A-24, A-25, B-29, C-69, C-54, C-87
Fighters	P-39, P-40, P-47, P-51, P-63

Lockheed Lightning P-38. WWII's "most useful and versatile fighter and medium attack bomber," per Lockheed.

horsepower of the engine, and how many engines the airplane had.

After the women pilots had flown ferrying missions for over a year (September 1942 to December 1943), they had accomplished instrument ratings, twin-engine missions and much flying time in a great variety of aircraft.

In June 1944 the 635 WASP on operational duty were classified as shown in the table. (Not all of the 635 WASP served in the Air Transport Command. Some of them served in the Training Command.)

"Upon graduation a WASP was given a Class I flying rating, the same rating as that of a male pilot. WASP on operational duty followed the regular pattern of on-the-job training and school attendance, and they advanced to higher ratings as flying civilian pilots.

"In the summer of 1944, 90 percent of the ATC women pilots were Class III pilots engaged almost solely in ferrying pursuits. It is well-known that these single-engine, fast planes are the most dangerous aircraft to fly and to ferry."

A requirement for Class III Pilot Pursuit classification was holding an instrument rating, which was possible to obtain at St. Joseph, Missouri, Instrument School. Some WASP, however, were able to obtain an instrument rating at their home base.

Barbara Erickson, Squadron leader of WASP 6th Ferrying Group at Long Beach, California is ready for take-off in a P-38. (Barbara Erickson collection)

P-38 Interceptor had "fastest climb and longest range of any fighter" in WWII per Lockheed Aircraft.

Gayle Bevis "take-two".

P-38 Lockheed Lightning

P-38, "The Lockheed Lightning," transition or training was given only to Class IV pilots, as they were required first to have medium bomber, twin-engine proficiency and to have qualified in pursuits.

The P-38 did have some unique flying problems, and was not regarded by some ferrying pilots as a desirable step-up or upgrade. However, if you were a pursuit pilot and had the opportunity to check out and fly the B-25 "Mitchell Bomber," the B-26 "Martin Marauder," A-20, or P-61 "Northrup Nighthawk Black Widow," you were an eligible Class IV pilot and could be checked out in the P-38. All pursuit airplanes at this period in history (1942-1945) had space for one pilot only, so the check-out consisted of studying many hours in the classroom and passing tests on the P-38 airplane and its operation, the cockpit check with an instructor beside you on the wing, then the solo and practice landings alone.

★★★

Less known than his flight across the Atlantic in 1927 was Charles Lindbergh's great contribution to improving the performance of the P-38 in WWII. Ferrying pilots were under orders to slowly break-in the plane engines (much like a car); thus, plane carburetors were being fouled with carbon buildup and would malfunction during combat action. Lindbergh flew the P-38 at a speed of about 1600 RPM, leaning out the fuel mixture at cruising altitude, then increasing prop pitch and manifold pressure before landing. This increased miles-per-gallon to one-third more, and did no damage to the engines. Needless to say, the ferrying pilots, combat pilots and, indeed, the U.S.A. as a whole owe more to Charles Lindbergh than appears in most history books.

★★★

Women pilots and their Ferrying Groups of the Air Transport Command who were Class IV Pursuit Pilots and qualified to fly the Lockheed Lightning P-38 were:

Nancy Batson	2nd F. Gp.
Lauretta Beaty	6th F. Gp.
Delphine Bohn	5th F. Gp.
Marion Carlstrom	5th F. Gp.
Helen Mary Clark	2nd F. Gp.
Mary Lou Colbert	21st F. Gp.
Lewise Coleman	6th F. Gp.
Iris Cummings	6th F. Gp.
Ruth Dailey	5th F. Gp.
Jane Emerson	5th F. Gp.
Barbara Erickson	6th F. Gp.
Betty Gillies	2nd F. Gp.
Kay Gott	5th F. Gp.
Byrd Granger	21st F. Gp.
Virginia Hill	6th F. Gp.
Vega Johnson	21st F. Gp.
Ross Kary	5th F. Gp.
Dorothy Kocher	6th F. Gp.
Katherine Loft	6th F. Gp.
Nancy Harkness Love	2nd F.Gp.,HQ
Helen McGilvery	2nd F. Gp.
Florene Miller	6th F. Gp.
Sidney Miller	5th F. Gp.
Nadine Ramsey	6th F. Gp.
Helen Richards	6th F. Gp.
Barbara Russell	21st F. Gp.
Evelyn Sharp	6th F. Gp.
Ruth Thompson	6th F. Gp.
Mary Trotman	21st F. Gp.
Catherine Vail	21st F. Gp.
Martha Wagenseil	21st F. Gp.

Douglas A-20A Army Attack Bomber

Virginia "Tex" Clair on a P-39. (Virginia Clair collection)

Women pilots had no opportunity to fly the P-38 at the 3rd Ferrying Division, Romulus, Michigan, near Detroit, but specialized in P-47 deliveries. Only four women in Wilmington, Delaware, 2nd Ferrying Division, qualified in P-38s. Helen Mary Clark verified that she went to Long Beach just for this transition.

P-39 Bell Aircobra

P-39 Bell Aircobra/P-63 Bell Kingcobra

The P-39 Bell Aircobra was manufactured at Niagara Falls, New York. The early models were powered by an 1150P Allison engine which was behind the pilot, and a 9-foot shaft turned the three blades of the propeller. P-39 Q was modified slightly to 1200P.

Air Transport Commander General Tunner related the following regarding P-39s.

"When we began experiencing a high degree of accidents in the deliveries of P-39s, most of them of such seriousness that the pilot was killed and the plane a complete washout, I put the cause down primarily as pilot failure. Sure, the P-39 was a hot ship, all right, but it was perfectly safe if it was flown according to specifications. The airplane needed speed, at least 150 mph, to maneuver. It had the glide angle of a brick, and before adequate baffles were placed in the tanks, gas sloshing around didn't help. Accidents occurred on the first turn after taking off, or the last turn coming in. The reason was obvious: on take-offs, pilots put the plane into steep bank before required speed; on landing, speed was reduced too much before the last turn. Any pilot who read 'specs' on the plane knew that it would go into a high-speed stall under such conditions. We were getting a lot of static from pilots who claimed the P-39 to be a flying coffin.

"There was a common feeling among many male pilots that the P-39s were 'suicide' planes; and that if one ferried them long enough, his chances of avoiding a fatal accident were slim.

"There were usually cumbersome tanks attached to these airplanes. In combat, these tanks were usually dropped before landing. When they were full, these heavier tanks made the P-39 almost unflightworthy, and were directly responsible for fatalities suffered by ferry pilots.

"Our women in the meantime had proved themselves as ferry pilots. They paid attention in class; they read the characteristics and specifications of the plane they

Women in Pursuit

3rd Ferrying Groups Marge Ketcham, Rene Boyd, Jary Johnson and Flight officer Story strike a pose on a flight line of a P-39 Air cobras destined for Russia (note red star). (Betty Shea collection, Betty married Rene.)

The Pursuit Song

Oh, pursuit ships, they don't bother me
Pursuit ships, they don't bother me
Just push on the throttle and up with the gear
We are the buzz girls without any fear
So start saying goodbye to them all
As into our coffins we crawl
They'll be no promotions this side of the ocean
The Bell Booby Trap kills them all
Kills them all, kills them all
The long and the short and the tall
Damn these instructors who taught us to fly
Damn these old death traps in which we must die
So start saying goodbye to your life
Cause Waspie, you'll ne'er be a wife
I've got a notion you'll not cross the ocean
The Bell Booby Trap kills them all.

(Ginny Hill collection)

were to fly before they flew it. The solution of the P-39 problem was a natural one; therefore, with no doubts whatsoever, I had a group of girls checked out on P-39s and assigned them to make P-39 deliveries.

"They had no trouble; none at all. And I had no more complaints from the men."

Eventually, the P-39 was modified at the Bell factory. Transition to the new model Bell, the P-63, was given to all P-39 qualified pilots at the Niagara Falls Bell factory, which included Kay Gott on July 1, 1944, Dorothy Kocher on October 3, 1944 and Gretchen Gorman on November 13, 1944.

P-39 (top) and
P-63 (bottom) in flight.

A P-40 Curtiss Kittyhawk awaits its pilot.

P-40 Curtiss Kittyhawk

Nearby at Buffalo, New York, was the Curtiss factory, where the P-40 "Kittyhawk" was produced.

Very often, the ferry pilots would be sent from a Newark, New Jersey, delivery to Niagara Falls for a Bell aircraft, or to Buffalo for a Curtiss P-40. It was possible to continue on the same set of orders from the original base until you ran out of deliveries.

P-47 Republic Thunderbolt

Nancy Batson was Commanding Officer of WASP squadron at the Republic P-47 factory in Evansville, Indiana.

WASP rotated in to Evansville for a four-week stint, and ferried new P-47s to Newark, New Jersey, port of embarkation for the European theater. It was a remarkable achievement that WASP were stationed at the factory and had the exclusive on P-47 deliveries.

"Between the Farmingdale, Long Island, Republic Thunderbolt P-47 factory and Newark, New Jersey, a quick shuttle trip was repeated by individual WASP as many as five times in a single day, involving repeated take-offs and landings, the times when accidents are most likely."

The Wilmington 2nd Ferrying Group organized such a successful operation of P-47s from factory to delivery that the entire Republic shuttle was run by the women pilots; even the pick-up C-60s were flown by WASP for the return from the delivery point.

WASP rotated in a four-week assignment to the Republic factory P-47 Thunderbolt at Evansville, Indiana, and at the Farmingdale plant on Long Island, New York.

From Farmingdale, it was the usual 30-minute flight to the overseas delivery

P-47 Republic Thunderbolts on their way to their destination.

point at Newark, New Jersey. All the pilots stationed at Farmingdale were women. They were picked up in Newark and returned to Farmingdale by a C-60 cargo transport. It was a very efficient and effective operation, completely planned and run by women!

The Evansville Republic site was a longer flight. WASP rotated there for four weeks duty also.

The East coast 2nd Ferry Group at Wilmington, Delaware, WASP and the 3rd Ferry Group at Romulus, Michigan, WASP rotated on this assignment that consisted of a fairly short hop, a tiring wait to return to the Farmingdale plant to bring another P-47 to Newark, and so on through the daylight hours.

★★★

Teresa James, an early WAFS, had the honor of delivering the 10,000th Thunderbolt, christened "Ten Grand", off the assembly line.

P-51 North American Mustang

Upon graduation from Palm Springs Pursuit School in March 1944, Kay Gott was one of the graduates ferried by TARFU to Long Beach, each pilot to pick up a P-51 to deliver to Newark, New Jersey.

In Atlanta, Georgia (refueling stop), many pilots RON'd. There on the airfield, aircraft were parked in neat long rows, one behind another. This was standard practice at USA bases everywhere, with devastating results at Pearl Harbor and war bases in the early war years.

That evening in Atlanta, Georgia, with Margaret Mitchell's book "Gone With the Wind" just published in 1939, and Peach Tree Street available, she could imagine what it was like in 1864 at the beginning of the Civil War.

Next morning after breakfast, weather check, clearance and telegram reporting off for Newark, Kay was not able to locate her P-51B. Upon walking down the flight line, she saw "Christmas tinsel" scattered all down the taxi ramps. The tail with the number markings of her assigned

P-51A with greenhouse canopy destined for China.

plane was chewed into aluminum bits! The propeller of the P-51 behind it had taken the tail off!

Indignant at having just graduated into pursuit ferrying and being unable to complete her first pursuit delivery, she asked for permission to take the undamaged airplane on to the Newark destination. The 'guilty' P-51 had to be inspected to be sure the propellers and the prop shaft were not damaged upon impact with Kay's P-51 tail. She had inspected the 'guilty' P-51 props earlier for nicks, as she was eager to take it on to Newark in place of her own plane. But after a few hours delay, she was sent back to Love Field, Dallas, on a commercial airline.

This 'guilty' P-51, however, was later assigned to another woman ferry pilot and taken to its destination for the European War Theater.

The guilty man, who did not look before he taxied forward to do the damage, was taken off pursuit ferrying.

Spare Parts

Kay wrote: "On my next flight through Atlanta, I inquired about 'my' P-51 to learn that parts had been 'pirated' to repair other P-51s. First to go were the tires."

In 1944, it was sometimes easier to get an entire new pursuit airplane than a part to repair one.

Logbook

Pilots were to note irregularities and dangerous flaws of the airplane in the Aircraft Logbook carried on that particular airplane.

Once qualified on an airplane, every pilot was supposed to remember all the details and peculiarities of that plane. Most women made their own specification notes — "spec" sheet — for each aircraft, procedures for take-off and landings, and in-flight settings. It was not uncommon in pursuit ferrying to fly three different types of fighters while on one set of orders from home base.

Cockpit/Cargo Space

The "hot" airplanes, the pursuits, the fighters — all single seaters — were not built for a tall person. Cockpit space for baggage was nonexistent. It was an advantage for a ferrying pilot not to be too tall, for the space beneath the seat could be used to stow some things when it was raised. Some tall pilots' heads nearly touched the canopy. It was also an advantage not to be too short. One or two women pursuit pilots (Vi Thurn for one) carried several cushions as part of her luggage in her parachute bag. She placed them behind her in the P-51 so she could reach the rudder pedals with her feet. Betty Gillies had wooden blocks built especially for her to help her reach the rudder pedals.

Women pilots were not alone in stature problems. Captain Robert E. Scott of P-40 fame in the China-Burma fight was an extremely tall man — over six feet. When interviewed, he reported that his crew chief modified his parachute pack so it fit in front of his chest rather than as the traditional seat pack.

Florene Miller put her skirt and dress clothes in the wing by unscrewing the compartment on the leading edge of the wing which was designed to contain ammunition.

Air Relief

Some distinctly funny problems occurred in flight in a single-seated fighter airplane: it was impossible to relieve yourself in the air! Women pilots were unable to use the men's relief tube, although in desperation some women did try. It was impossible to hit the receptacle. Let's face it, women's plumbing is just not the same! After near disasters with internal problems, the women found it highly desirable to skip some food and drink items, as well as all carbonated beverages and beans, since some pursuit flights with external fuel tanks were four hours in the air.

Cochran wrote: "The fighter planes have so-called service tubes for the men, but these don't work for the female of the species. I tried various expedients, but most workable of all was to dehydrate myself before the race and then fight nature. During the later WASP days, the laboratories at Wright Field worked out an apparatus to meet this situation."

Unfortunately for the women ferrying the fighters in World War II, the developments for women's comfort never reached down the echelon from the Wright Field laboratories to the women who were flying.

Helen Mary Clark WAFS, Teresa James WAFS, Irene Gregory WASP 43W-5 attend OTS in Orlando, Florida. (Teresa James collection)

Margaret Kerr WASP 43W-2, Marge Logan WASP 43W-6 in informal WASP uniform, Helen Richey ATA/WASP, Irene Gregory 43W-5; WAFS Helen Mary Clark in WASP uniform; WAFS, Teresa James in WAFS uniform OTS Orlando, Florida. (Teresa James collection)

Officer Training School

All pursuit ferrying ceased for one month for each woman as the WASP were rotated to Officer Training School (OTS) to learn to be an officer in the US Army. This school was held at Orlando, Florida, and women were rotated there in order of seniority in the WAFS and WASP programs.

Most of the women pilots considered that month at OTS a waste of flying time, since WASP served under Civil Service and were dismissed as civilians. Veterans' status was only awarded some 35 years later. However, 460 women pilots went through that training program to become Army Air Force Officers before the school was terminated for women.

★★★

On a Daytona Beach outing, Evelyn Greenblatt rented a car, and five WASP spent a day on the beach. They got burned on the white sand and had to sit in a hot humid classroom, made bearable only by wearing rayon pajamas under wool slacks.

War-weary

War-weary airplanes sometimes fell apart as they were being flown to depots. On occasion, a fighter plane from a training school was delivered to a junk yard. At the end of 1944 to ferry a war-weary airplane was very "chancy," not highly regarded by any ferry pilot as a desirable job. Moving schools to a different location was also a dubious honor, as some aircraft had been beaten up by over-controlling students, dropped into landing fields from too high altitudes, and had worn cables. All things considered, there were too few qualified, conscientious mechanics, and too little time for them to do thorough checks of each airplane. It was up to ferry pilots and all

pilots to note any deficiencies of any airplane they flew.

Red-lines and X-rated aircraft became common at the end of 1944. This meant danger to any pilot. Yet the aircraft was slated to be moved! And move aircraft the WASP did!

Combat pilots who completed their 50 missions flying pursuit in the war were often sent to ferrying bases in the United States to take on the "less hazardous" job of delivering airplanes, not to mention "would-be" combat pilots in the USA relegated to domestic deliveries of pursuit airplanes!

"The most interesting conclusion is that throughout their career the women ferry pilots concentrated upon types of ferrying essentially more hazardous than done by their male colleagues. So far as it is known there was no tendency to complain about this; if anything, at least in the pursuit period, it was a matter of pride."

The Beginning of the End

On October 15, 1944, the last class closed for women students in pursuit training; and General 'Hap' Arnold announced that on December 20, 1944 the Women Airforce Service Pilots would be dismissed.

There had been 19 classes in pursuit flying training for the women pilots in ferrying. It was unbelievable to the women flying pursuits then, at the height of their usefulness to the government and to the war effort, that they should be "fired."

The "ninety nines, inc., International Women Pilots," a women's flying group, originally established by 99 women with Amelia Earhart, wrote Congress to protest the WASP dismissal before the war ended. Some of the women in pursuit ferrying wrote letters to Congress. Most, like the author, did not really believe that Congress would waste the taxpayers' money, after all they had spent for training; and WASP had a marvelous record of deliveries of pursuits.

★★★

"We did not believe that they would release us, and tell us to go home: 'Goodbye, we don't need you anymore!'

At that time, there was a crisis — a pile-up of pursuit planes at factories needed to be moved. The Air Transport Command asked for an exception so that the women who wanted to could continue. It was deemed by Congress that there was a sufficiency of pilots, and the women were taking jobs from male pilots.

Some women pilots volunteered to fly for no compensation; other women asked to continue to fly for a dollar a year as some valuable citizens did during those war years.

To no avail! WASP, in spite of an exemplary record and their great desire to to continue to help the war effort, were not allowed to see it through! They were out of ferrying pursuits December 20, 1944.

Last Deliveries

The last flight delivery from the Hensley Field North American factory by a WASP of a P-51 Mustang fighter was in a P-51, called an F-6K, to Meridian, Mississippi. It was December 13, 1944 and deactivation of the WASP was only seven days away. This flight was made by Kay Gott from the 5th Ferrying Group at Dallas, Texas.

★★★

When the WASP deactivation date came, December 20, 1944, and the women could no longer fly for the government, there was great anticipation as to who would get the very last flight. On December 19, 1944, Anna Flynn was the lucky WASP from 2nd F.Gp. Wilmington. She drew the match for the honor of the last delivery. She still has the match-stick in her scrapbook, with her memories.

Early Love Field 1943,. P.R. photo with WASP Squadron leader Florene Miller pointing the route to Betty Jane Bachman, Avanell Pinkley, Catherine Vail and Betty Whitlow with Fran Dias standing. Beechcraft twin engine in background. (Cappy Vail collection)

P-39

Pre-Pursuit School Transition
Palm Springs, California
November 1943

Barbara Jane Erickson, WAFS Long Beach, CA

Barbara Towne, WAFS Long Beach, CA

Class 43-1
Pursuit Transition School
Palm Springs, California
Ferrying Division, Air Transport Command
December 1943

Iris Cummings, Redondo Beach, CA

Ruth Thompson, San Diego, CA

Class 43-1 & 2
Pursuit Transition School
Palm Springs, California
Ferrying Division, Air Transport Command
December 1943

front row
Betty Jane Bachman, Dallas, TX
Lewise Coleman, Oakland, CA
Florene Miller, WAFS
Sidney Miller, Mineral Wells, TX
Gertrude Meserve, WAFS
Helen Richards, WAFS Boise, ID
second row
Carole Fillmore, Menlo Park, CA
Dorothy F. Scott, Spokane, WA

(Fred Adams collection)

Class 44-3
Pursuit Transition School
Palm Springs, California
Ferrying Division, Air Transport Command
January 1944

Betty Whitlow, Tulsa, OK
Betty Jane Bachman, Dallas, TX
Norma Jane Emerson, Kansas City, KS
Sidney Miller, Mineral Wells, TX
Catherine Vail, Washington, DC
(Cappy Vail collection)

Class 44-3
Pursuit Transition School
Palm Springs, California
Ferrying Division, Air Transport Command
January 1944

Nancy Batson, WAFS
Birmingham, AL

Helen McGilvery, WAFS

Class 44-4
February 1944

Mary Catherine "Jary" Johnson, Orchard Park, NY
Barbara Donahue, New York, NY
Emily Hiester, Reading, PA
Jane Straughan, Washington, DC
(Fred Adams collection)

Class 44-5
March 1944

Ruth Dailey, Grapeland, TX
Byrd Howell Granger
Kathrin Gott, Nampa, Idaho
Marjorie Jane Ketcham, Washington, DC
Ross Kary, Kary, SD
(Kay Gott collection)

Class 44-6
March 1944

Virginia Alleman, Washington, DC Ruth Grimm Trees, Indianapolis, IN Mary T. Trotman, Chevy Chase, MD

Last class at Palm Springs Pursuit School

Class 44-7
April 1944

Mary Lou Colbert, Washington, DC
(Collection of WASP Historian Mary Anna Wyell)

First class at Brownsville, Texas Pursuit School

Class 44-8 Flight A
May 1944

Vega Johnson, New York, NY
Betty C. Tackaberry, Honolulu, HI
Ellen H. Gery, Reading, PA

Claire G. Callaghan, Washington, DC
Ruth E. Franckling, Woodstock, NY
Martha Wagenseil, Pittsburgh, PA
(Fred Adams collection)

Class 44-8 Flight B
May 1944

Margaret A. Hamilton, Enid, OK
Margaret E. Kerr, Ada, OK
Gretchen J. Gorman, Bottineuau, ND
Nadine B. Ramsey, Wichita, KS
Cornelia Y. Colby, Jacksonville, FL

Barbara Willis, Bound Brook, NJ
Mary E. Darling, Longmeadow, MA
not pictured: Florence P. Lawler, Fort Worth, TX
Barbara Russell, San Mateo, CA
(Fred Adams collection)

Class 44-9 Flight A
May 1944

Virginia Clair
Virginia Harris*
Betty Naffz*
Dorothy R. Colburn
Ruby Mullins*

Jean Mohrman*
Dorothy M. Nichols
Isabel Madison
Rita J. Moynahan
Marjorie Gray*

Elizabeth C. Shea
Harriet Urban*
*did not complete this course
(Betty Shea collection)

Class 44-9 Flight B
May 15, 1944

Lauretta Beaty, Burbank, CA
Janet Zuchowski, Newburgh, NY
Betty Archibald, Chicago, IL
Juanita Bolish, Los Angeles, CA
Evelyn Tomlinson, San Diego, CA

Janice Tate, Dallas, TX
not pictured:
Violet Thurn, Bowdle, SD
Rena D. Wilkes
(Fred Adams collection)

Class 44-10 Flight A
June 1, 1944

Katherine S. Loft, Miami, FL
Jean Trench,* Staten Island, NY
Helen I. Fremd, Canal Point, FL

Marion V. Carlstrom, Cowdry, CO
Virginia Luttrell, Hollywood, CA
Dorothy E. Kocher, Woodburn, OR

Hazel W. Pracht,* Tamarack, MN
* did not complete the course
(Katherine Loft collection)

Class 44-10 Flight B
June 1, 1944

Esther D. Poole, Houston, TX
Virgnia Hill, Seattle, WA
Lillian Conner, Pomona, CA

Genevieve Brown,* Seattle, WA
Grace P. Birge, Cleveland, OH
(Virginia Hill collection)

Class 44-11 Flight A
June 16, 1944

Eleanor E. Thompson, Spencer, NC
Isabel M. Steiner, Palo Alto, CA
Kittie L. Leaming, Upper Darby, PA

Mary C. Wilson, Springfield, OH
Madge Rutherford, Indianapolis, IN
(Madge Rutherford collection)

Class 44-11 Flight B
June 16, 1944

Ruth E. Anderson, Seattle, WA
Rosa Lea Fullwood, McAllen, TX
(Fred Adams collection)

Class 44-12 Flight A
June 30, 1944

Dorothy E. Webb, Bakersfield, CA
Nancy L. Featherhoff, St. Louis, MO
Ruth Lindley, San Diego, CA

Enid C. Fisher, Portland, OR
Patricia A. Dickerson, Lafayette, IN

not pictured:
Gayle D. Bevis   Alice Jean May
(Enid Fisher collection, courtesy of Terrill M. Aitken)

Class 44-12 Flight B
June 1944

Mary Rosso, Kalamazoo, MI
Josephine A. Pitz, Manitowoc, WI
Celia M. Hunter, Everett, WA

Lana B. Cusack, Boise, ID
Helen A. Turner, Cairo, NB
(Lana Cusak and Celia Hunter collection)

Class 44-13 Flight A
July 15, 1944

Margaret L. Castle, Muroc, CA
Marcie Jo Myers, Dallas, TX
Kathryn L. Bernheim, Hewlett, LI, NY

Virginia L. Jowell, Frankston, TX
Jean Landis, Long Beach, CA
(Jo Myers collection)

Class 44-13 Flight B
July 15, 1944

Rosalie L. Grohman, Saginaw, MI
Evelyn L. Trammell, Atlanta, GA
Edith D. Truax, Los Angeles, CA

Betty Scantland, Indianapolis, IN
(Fred Adams collection)

Class 44-14 Flight A
August 1, 1944

Lillian M. Conner, Pomona, CA
Anna L. Flynn, Natick, MA
(Fred Adams collection)

Class 44-14 Flight B
August 1, 1944

Marian G. Mann, Canton, NC
Lola C. Perkins, Los Angeles, CA

Gwendolyne E. Cowart, Atlanta, GA
Mitchell I. Long, Rome, GA
(Fred Adams collection)

Class 44-15 Flight A
August 15, 1944

Marianne I. Beard, Milwaukee, WI
(Fred Adams collection)

Class 44-15 Flight B
August 15, 1944

Ruth Adams, Cleveland, OH
Thelma N. Harris, Chanute, KS
Vivian C. Cadman, Fullerton, CA

Sarah E. Pearce, Quincy, FL
(Ruth Adams collection)

Class 44-16 Flight A
August 31, 1944

Rebecca H. Edwards, Yazoo, MS
(Fred Adams collection)

Class 44-16 Flight B
August 31, 1944

Margaret Wendelin,* Seattle, WA
Margery Taylor,* Wellesley, MA
Helen M. Schaefer, Cincinatti, OH
*did not complete course
(Fred Adams collection)

Class 44-17 Flight A
September 15, 1944

Dorothy Hopkins, Brooklawn, NY
Virginia B. Crinklaw, Newman, CA
(Fred Adams collection)

Class 44-17 Flight B
October 1, 1944

Gertrude V. Tompkins, Summit, NJ
Nelle L. Carmody, Los Angeles, CA

Helen T. Abell, Springfield, IL
(Fred Adams collection)

Class 44-18 Flight A
October 1, 1944

Eleanor A. Alexander, Santa Monica, CA
Avanell Pinkley,* New York, NY

Nancy L. Baker, White Plains, NY
*did not complete the course

(Fred Adams collection)

Class 44-18 Flight B
October 1, 1944

Hazel Ying Lee, Forest Hills, NY
Frances M. Snyder, Rupert, ID
Joan M. Trebtoske, Portsmouth, NH
Sylvia A. Dahmes, Redwood Falls, MN

Helen F. Barrick, Sedalia, MO
Maurine Miller, Evanston, IL
Jane S. Scott, Pasadena, CA
(Sylvia Dahmes collection)

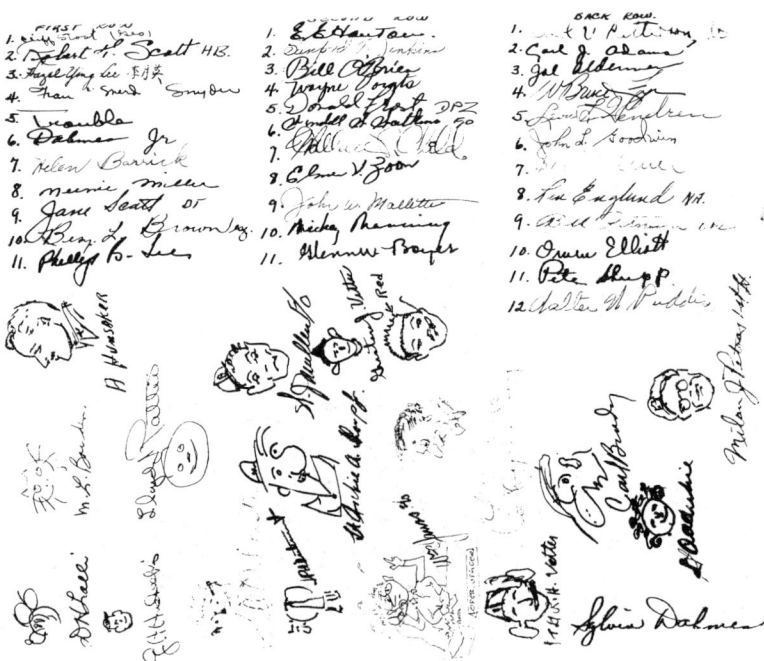

Class 44-19 Flight A
October 15, 1944

Virginia L. Sweet, Schenectady, NY
Irene G. Davis, Philadelphia, PA
(Irene Gregory collection)

Class 44-19 Flight B
October 15, 1944

Jill McCormick, Greensburg, PA
Louise Bowden, Elmer, NJ
not pictured: Alice E. Lovejoy - killed
in training at Brownsville, TX.
(Jill McCormick collection)

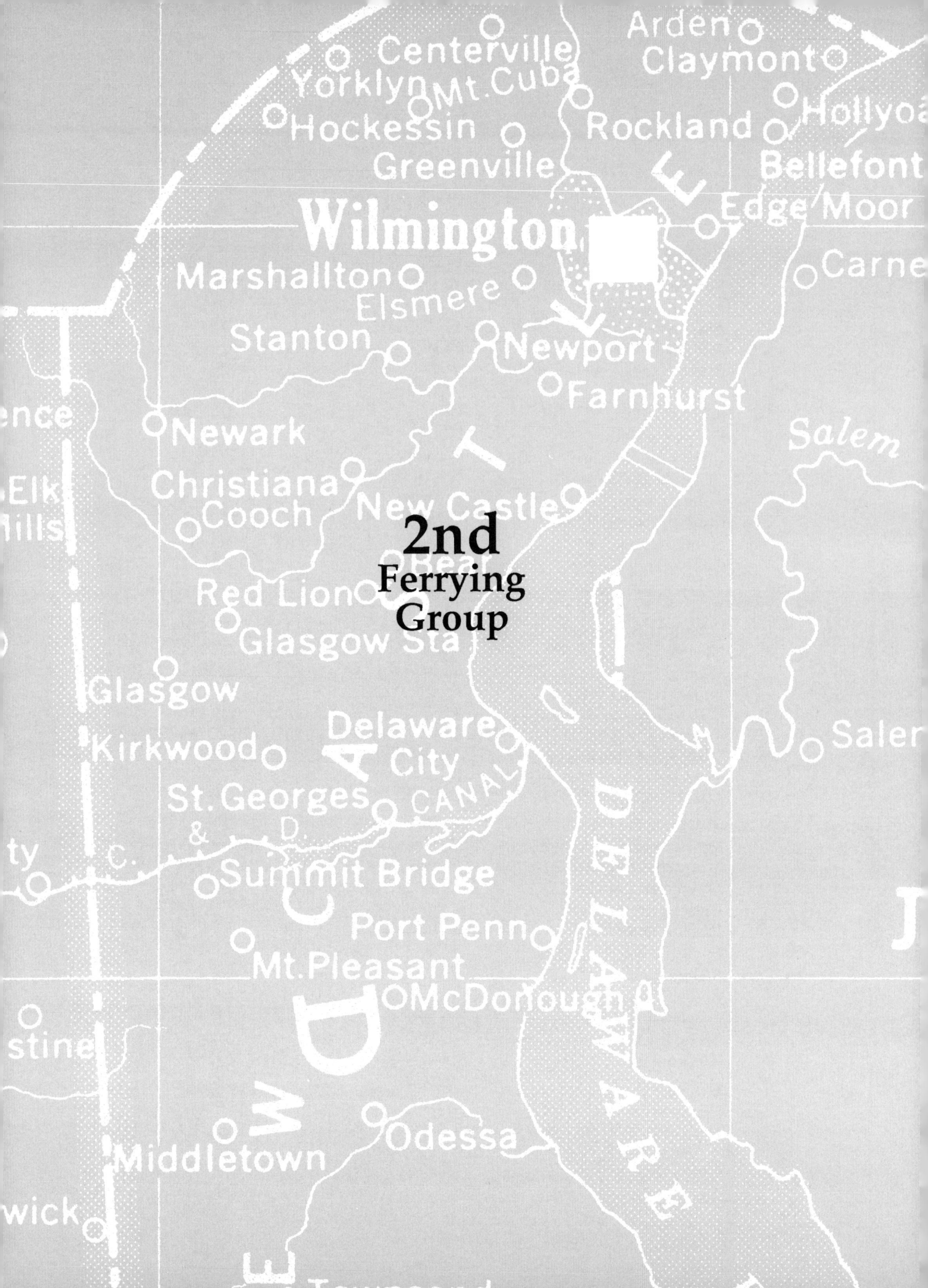

2nd Ferrrying Group
Air Transport Command, Wilmington, Delaware
New Castle Army Air Base

Name	WASP Class	Pursuit School	Graduation Date
Adams, Ruth	44-W-2	44-15B	8/15/44
Alleman, Virginia A.	43-W-2	44-6	3/44
Anderson, Ruth E.	43-W-5	44-11B	6/16/44
Baker, Nancy Lee	43-W-4	44-18A	10/1/44
Batson, Nancy	WAFS	44-3	1/44
Beard, Marianne	43-W-5	44-15A	8/15/44
Bernheim, Kathryn "Sis"	WAFS	44-13A	7/15/44
Clair, Virginia "Tex"	43-W-4	44-9A	5/15/44
Clark, Helen Mary	WAFS	—	
Colburn, Dorothy R.	43-W-4	44-9A	5/15/44
Cowart, Gwendolyne E. "Gwen"	43-W-4	44-14B	8/1/44
Flynn, Anna L.	43-W-6	44-14A	8/1/44
Gillies, Betty Huyler (Squadron Leader)	WAFS	—	
Gregory, Irene G.	43-W-5	44-19A	10/15/44
Grohman, Rosalie L.	43-W-4	44-13B	7/15/44
Hiester, Emily	43-W-2	44-4	2/21/44
Hunter, Celia M.	43-W-5	44-12B	6/30/44
James, Teresa D.	WAFS	—	
Jowell, Virginia "Virgie Lee"	43-W-4	44-13A	7/15/44
Long, Mitchell I. "Mitch"	43-W-7	44-14B	8/1/44
Love, Nancy Harkness Executive Air Transport Command	WAFS	—	
Mann, Marian G.	43-W-7	44-14B	8/1/44
McCormick, Jill	43-W-5	44-19B	10/15/44
McGilvery, Helen	WAFS	44-3	1/44
Meserve, Gertrude "Gert"	WAFS	44-2	1/44
Moynahan, Rita Joan	43-W-2	44-9A	5/15/44
Pearce, Sarah E. "Liz"	43-W-4	44-15B	8/15/44
Pitz, Josephine "Jo"	43-W-5	44-12B	6/30/44
Poole, Esther D.	43-W-3	44-10B	6/1/44
Richey, Helen	ATA 43-W-5	—	
Rosso, Mary M.	43-W-4	44-12B	6/30/44
Straughan, Jane S.	43-W-1	44-4	2/21/44
Trees, Ruth Grimm	43-W-2	44-6	3/44
Wilson, Mary C. "Sleepy"	43-W-4	44-11A	6/16/44
Zuchowski, Janet J.	43-W-4	44-9B	5/15/44

In the 2nd Ferrying WASP squadron, a May 25, 1944 report states that 19 WASP pilots qualified in pursuits at Wilmington, and almost all of them got their time in delivering P-47s from Farmingdale, New York, to Newark, New Jersey, a 30-minute flight. During the year and a half the WASP were assigned to the Republic factory they delivered 2000 P-47's from the factory. The average number of planes delivered was two or three, with a logged time of an hour and a half per day. If the weather was perfect, and a plane was available in Newark to take the pilot back, everything worked like clockwork.
A WASP might accumulate 45 hours in a 30-day month.

Marx writes "The process would be exhausting."

Each woman pilot was unique, and each woman pilot's story is different. There is really no "typical" pattern. The women pursuit pilots based at Wilmington will have similar stories to tell because the ferrying task job at Wilmington was shared by the women there who flew pursuit airplanes.

Nancy Harkness Love, Betty Gillies and Helen Mary Clark were Executives of the WASP Squadron at Wilmington, and a "shuttle run" was made possible by the astute planning of these women pilots. They arranged with the Commanding Officer of the base for a "pick-up" airplane, a cargo C-60, to shuttle these women pilots back to the Long Island factory after delivering their P-47s to Newark, New Jersey. Gertrude Meserve was the "pick-up" pilot Jo Pitz remembers.

It was possible to deliver up to six airplanes in one day depending upon weather conditions, factory readiness of the plane, and the expert timing of the pick-up aircraft for their return. The women lived on Long Island at the Bayshore and Huntington hotels, and were ready to deliver one to six airplanes a day, Sundays included.

Betty Huyler Gillies, Squadron Leader, 2nd Ferrying Group WASP, New Castle Army Air Base, Wilmington, Delaware. (Betty Gillies collection)

Betty Gillies

Betty Huyler Gillies was a very experienced pioneer pilot, one who was already licensed in 1929, and who, with Amelia Earhart, help found the "ninety-nines" organization for women pilots.

Betty was Squadron Leader of the WASP at the 2nd Ferrying Division at Wilmington, Delaware, and the first original WAFS recruited by Nancy Love.

These notations from Betty's logbook, showing a "typical work day", were published in the P-38 bulletin Vol. 2, No. 1, January 1989.

14 July 1944: Arrived Middletown PA at 1740 to pick up P-38F 42-12587. Ship not ready. RON Harrisburg.

20 July 1944: Out at field 0830 and there the rest of the day. Ship still not ready. Could not get fuel flow from the left belly tank. Compass NG. Left mag on right engine NG. RON Harrisburg again.

21 July 1944: Out at field 0800. At 1500 plane finally declared ready to go. Took off at 1550 cleared for Scott Field (near St. Louis). After running off belly tank for 15 min., the fuel pressure of the left engine went down and the engine quit. Landed at Patterson instead of Scott. Lineman at Patterson found large wad of masking tape which had been used to plug vent on left tank. Filed for Scott Field. Taxiing by the tower, I was informed that the gasoline was siphoning out the overflow on the right tank. Out to remove cap but gas continued to siphon until tank was empty, as fire trucks stood by. Refueled to go again (as no one seemed to know why the tank had emptied) and amperage shot up to over 100 and stayed there. RON Patterson Field.

22 July 1944: New battery installed. Took off for Tulsa OK at 1020. Everything seemed OK except that the front tank kept filling up and I had to run off it for 20 minutes of each hour. 3 hours 20 minutes to Tulsa. Bad weather ahead. RON Tulsa.

23 July 1944: Off at 0955 for Albuquerque, NM. On this leg the amps started going up again. Turned everything off but had to put it back on to land. By the time I had circled the field and landed, the battery was shot to hell. Had another battery installed and took off for Van Nuys, CA at 1430. On T.O. the turbo on right engine refused to operate. Could only get 35 in. for T.O. and 25 in. at 12,000. Some instrument weather en route. Landed at Van Nuys 1745. Ship delivered OK. Whew!

Betty Gillies related the following in a telephone conversation October 3, 1991:

The early WAFS who arrived at the 2nd Ferrying Division Air Transport Command at the New Castle Army Air Base, Wilmington, Delaware on September 1942 had one month taking military courses before beginning to ferry primary trainers (Fairchild PT-19) and the then Basic Trainers (Vultee BT-13 and 15s). Those 28 women all had at least 500 hours flying time prior to entering the Nancy Love program.

"New Castle Commanding Officer Baker was great! They wanted us because they had more planes to move than they could handle! The women were glad to be there. There was no struggle; they got along."

★★★

Transition was not set up at Wilmington like it was at Long Beach. There were no nearby aircraft factories, no airplanes just coming through. They never saw a Mustang P-51.

Betty went to Long Beach to learn to fly the P-63, P-61 and the P-38. Betty, Helen Mary Clark, Teresa James, Nancy Love and Barbara Erickson all flew pursuits before pursuit school was established, some ten months after the WAFS began delivering the primary trainers.

Betty insisted these were the BEST times! Only the most memorable, the most outstanding time of our lives — just the best years. Great fun, much laughter, and camaraderie. Not just great achievement, but a sense of true worth, proving women were capable — even more than they ever dreamed — and dream they did!

Jane Straughan

Kay: What did you think about your time in the WASP?

Jane: I enjoyed it very much. I was very thankful for the training we had, and I was very happy that I was at Wilmington.

K: Was it the high point in your life?

J: Yes, the high point that's different. It was the most unusual part of my life. I had no idea I would fly a fighter, ever.

The only time I was a little afraid was when I took a P-47 from Republic and they weren't flight-tested much. I had smoke come up into the cabin and I took it back in. I did damage the landing gear because the check pilot there thought I was falling and they followed me in and I was jittery. I dropped it a little, not too bad. It happened to be one that was going to Russia and that made it worse, because the Russian pilot was there. He couldn't talk to us but he kept telling me, through whoever was resident then, "off goes his head." What he was saying was that this airplane would be out of sequence and would be about two weeks late getting there.

K: Were you forbidden to talk to him?

J: No, we weren't forbidden; but he was forbidden to talk to us, so he met with us very seldom. You would see him but you wouldn't get into contact.

Jane S. Straughan

Jane Straughan in front of an AT-6. Note parachute over her shoulder and map pocket on her knee. (Smithsonian archives)

Helen Richey

The only women who had succeeded in aviation prior to 1942 were winners in races, holders of "firsts" in aviation, and some few taught at flying schools. One exception was Helen Richey who flew with Pennsylvania Airlines after air racing, and she flew in Britain, ferrying airplanes with the British ATA. She returned to join the Women Airforce Service Pilot program, graduated from Sweetwater (record time), was based in Wilmington, Delaware with the 2nd Ferrying Group, and went to the 33rd Ferrying Group ferrying B-25s out of Kansas City. She is in the last photograph at WASP Squadron, Wilmington, Delaware, when the WASP were disbanded.

Gertrude Meserve

"I will try to give you some additional information on why some girls went to Pursuit School at Palm Springs, California, and other girls didn't."

"Pursuit School had not opened up for the first three girls who checked out in the P-47 at Wilmington, Delaware. That included Betty Gillies, Helen Mary Clark and Teresa James."

"I made my original P-47 solo flight at Wilmington, Delaware, and then went to Palm Springs where I again checked out in the P-47, P-39, P-40 and the P-51. At the conclusion of pursuit school in December 1943, I ferried a P-51 from Long Beach, California, to Buffalo, New York."

"During our training at Palm Springs, Dorothy Scott was killed with her instructor in a mid-air collision on her approach to the airport. Helen Richards, Nancy Batson and I had the privilege to escort her body home to the Los Angeles area."

"I received my Class IV rating, but did not check out in the P-38. I had no desire to fly war-weary P-38s that had come back from combat."

Dorothy Colburn

After graduating from Pursuit School Class 44-9A at Brownsville, Texas, on May 18, 1944, Dorothy was stationed at "NCAAB 552nd BTU 2nd FG, FD, ATC" in Wilmington, Delaware.

She went to Instrument School at St. Joseph, Missouri, and OTS at Orlando, Florida, also in 1944.

Gert Meserve checks the view from a P-47 Republic Thunderbolt with bubble canopy. (Gertrude Meserve collection)

Dorothy Colburn keeps warm in winter garb issue. (Dorothy Colburn collection)

One-month assignments to both the Farmingdale, New York, and Evansville, Indiana, Republic factories found Dorothy delivering one to three P-47s a day from these locations to the Newark, New Jersey, port of embarkation.

Dorothy qualified in the P-39, P-40, P-47 and P-51, but also delivered AT-6s, PT-26s, RA-24Bs, B-25s and B-26s.

Esther D. Poole

Esther attended pilot training in WASP Class 43-W-3, at Sweetwater, graduating on 3 July 1943. She was assigned to the 2nd Ferrying Group at Wilmington, Delaware, and took transition to heavier aircraft since she was delivering training airplanes. As she progressed into heavier aircraft, she flew a DC-3 to Montreal, Canada. She got her Instrument Rating at St. Joseph, Missouri. She was never sent for Officer's Training in Orlando, so she spent the entire seven months ferrying fighters after graduating from pursuit school.

In the early years, she recalls a memorable flight of four WASP delivering slow Stinson hospital ships to Edmunton, Canada. Esther was flight leader, and she vividly remembers the split hospital beds on the right side of the plane and across the back.

In May, she was sent to Brownsville Pursuit School. Her classmates included Virginia Hill (6th FG), Grace Birge (3rd FG) and Lillian Conner (3rd FG). These four women graduated June 1, 1944 from pursuit school. After 11 months ferrying training planes, successfully completing instrument school, and being Class III pilots they were now Class III-Pursuit qualified. Esther returned to the 2nd Ferrying Division at Wilmington.

She never forgot one incident at Pursuit School. "In a P-39, the air scoop flew off on take-off, and all the instruments went 'haywire.' The tower screamed 'What are

Esther D. Poole

your intentions?' Too busy to say anything, I got off the ground and headed for Mexico, just over the border. I thought about bailing out; but someone had switched parachutes and this one was too large. The fire engines and the ambulance came out. I got it down, and they grounded me for three days. After that I was assigned to shoot landings. When I got back to Wilmington after graduating, I told Betty Gillies, our squadron Exec., **'Never put me on a P-39 delivery!'"**

The airplanes she flew, and the order in which she flew them at pursuit school were: AT-6s, P-39s, P-40s, P-51s and P-47s. Esther often delivered five P-47s in one day. Occassionally she would take a P-51 back from Newark to Evansville.

She served seven months ferrying fighters.

✯✯✯

Waiting for the weather to clear was often a problem. Esther recalls being weathered in most at Pittsburgh, Pennsylvania. Those boring times of waiting were devoted to catching up on sleep and writing letters. Flying was everything!

Mary C. Wilson

Mary Wilson

"I was in the 43-W-4 class at Sweetwater and we graduated August 7, 1943. I was assigned to 2nd Ferrying Division, ATC until WASP were disbanded. On April 8, 1944 I went to Instrument School at

At the P-47 Factory in Farmingdale, N.Y. Adorning jeep are Jo Pitz, Ruth Anderson, Nancy Batson, Celia Hunter, Mary Rosso, **Gertie** Meserve, "Sleepy" Wilson and Esther Poole. (Celia Hunter collection)

St. Joseph, Missouri where we trained on C-47s and C-49s. On May 16, 1944, I went to Pursuit School at Brownsville, Texas. I ferried P-47s from Republic Aircraft to Newark, New Jersey airport July 10, 1944 to August 14, 1944 while at Farmingdale, New York."

"The reason that I was not in the "last day" picture was that I had leave time and returned home before the others did."

WASP Anna Flynn checks the cockpit after delivering a P-51D. (Anna Flynn collection)

Anna L. Flynn

Anna Flynn

Anna Flynn received her wings from William T. Piper upon graduation of class 43-W-6 at Sweetwater on October 9, 1943. Mr. Piper was honored at this graduation ceremony as one of the major contributors to aviation. Anna was one of five WASP in pursuit, out of a total of 18 WASP who were former Piper employees and known as "The Piper Girls".

Anna won the luck of the draw to fly the last official WASP delivery on December 19, 1944, just one day prior to the congressional disbanding of the WASP.

William T. Piper with Jacqueline Cochran, Mrs. Barton K. Yount, wife of the Commanding General AAF Training Command.

Piper Alumnae at Sweetwater. All the WASPs in this picture, taken about July 1943, did some of their earlier flying at Piper Aircraft Corporation in Lock Haven, Pennsylvania. Jean Hoopes, Marcia Courtney Bellasai, Lauretta Beaty, Mildred "Duke" Caldwell, Libby Gardner, Dorothy "Rusty" McLean, Anna Flynn, Lola Perkins, Jean Bothwick, Alice Lovejoy, Anne Shields, Julie Ledbetter, Nancy Baker, Jeanne d'Ambly, Jean Trench and unidentified WASP Shirley Condit and Ruth Carter. WASP Newsletter, November 1983. (WASP Historian Collection)

Jo Pitz

Jo Pitz delivered 142 P-47s! She was in pursuits for only 166 days!

Was this accomplishment a figment of her imagination? Was she possessed of magical powers? Did she have amazing endurance? Was the "Fifinella" gremlin helping her by riding on her shoulder? Just how was this record possible?

Josephine Pitz

July 1944 Detachment of WASP at Republic Factory, Evansville, Indiana. On the ground, Major Walker, Commanding Officer; Nancy Batson, Jane Straughan, Jary Johnson, Emily Hiester, Virgie Jowell, Anna Flynn, Betty Scantland. On the wing, Rita Moynahan, Virginia Alleman, Helen McGilvery, Esther Poole, Pat Lawler, Kathryn "Sis" Bernheim. Back row, Mitch Long, Janet Zuchowski. (Betty Gillies collection)

The delivery of P-47s, the Republic Thunderbolt, the heaviest fighter the United States made during World War II, was flown from the factory entirely by women — not just one woman, Jo Pitz, but by all the women pilots in the Air Transport Command stationed at Wilmington, Delaware with the 2nd Ferrying Division, and at the 3rd Ferrying Group in Romulus, Michigan, who qualified and were eligible to fly pursuits.

Gertrude Meserve, Celia Hunter, Ruth Anderson and Jo Pitz look skyward after delivering P-47s from the Republic factory at Farmingdale, N.Y. (Celia Hunter collection)

Farmingdale WASP Nancy Batson, Jo Pitz, Ruth Anderson, Esther Poole, Mary Rosso, Celia Hunter, "Sleepy" Wilson and Gert Meserve flew planes (note British "bulls-eye" ID on wing) to Newark, N. J., for pickling. (Celia Hunter collection)

Through the examination of the flying record of Jo Pitz, a clearer picture emerges of the jobs of the women pilots stationed at Wilmington who were flying pursuit aircraft. Each base where women pilots were stationed was different, and the women's lives were not the same, nor the jobs the same, from one base to the other.

Jo spent six months training at Sweetwater, Texas, in primary, basic and advanced type training airplanes. She graduated from this school on September 11, 1943 with class 43-W-5, the fifth class of women pilots trained under the direction of Jacqueline Cochran. Also sent to Wilmington base from that class and eventually into pursuit flying were Ruth Anderson, Marianne Beard, Irene Gregory, Celia Hunter, Jill McCormick and Helen Richey.

After the 43-W-5 graduation in September 1943, Jo ferried aircraft out of the Wilmington base for the next eight-and-one-half months. Finally, on June 1, 1944, she was sent to Brownsville, Texas for one month's training in P-40, P-51, P-39 and P-47 pursuit aircraft. She delivered a P-63 to Great Falls, Montana for the Russian lend-lease program, and flew all of the single-engine fighters during her tenure as WASP.

In the 166 days Jo was eligible to fly pursuits, to deliver 142 Republic Thunderbolt P-47s was a remarkable feat! But she was not alone in this.

Some early Wilmington-based WAFS, such as Teresa James, obtained pursuit qualification at the factories; and some WASP and WAFS were successful in obtaining qualification at their bases. Gertrude Meserve, Wilmington WAFS, attended the second group in pursuit class at Palm Springs, January 1944. Helen Richey, an experienced pilot who served with the Air Transport Auxiliary in Britain and returned to the USA, was accelerated

Jo Pitz holds the prop of a P-47 (note "777" chalked on blade). (Jo Pitz collection)

through Sweetwater training, and graduated with that 43-W-5 class. Helen was sent to the Wilmington base, and flew the USA pursuits although she did not attend pursuit school in the United States.

Author's note: Her fascinating history in aviation has been written in "Propeller Annie" by Glen Kerfoot.

How was Jo Pitz able to do so many flights in that short time?

During the 22 months that Jo served, she never took a leave to visit her home in Wisconsin. She was eager to fly, and her achievement shows it.

An assignment to OTS came after she completed Pursuit School; but that school at Orlando, Florida closed to WASP on September 29, 1944, so she was spared a month sitting in a classroom learning to become an officer. She spent that month flying instead.

Jo Pitz and the Wilmington WASP in pursuits were able to deliver record numbers of P-47s because the Farmingdale P-47 factory was a short distance by air from Long Island to New Jersey. This was a 30-minute flight; but it could be longer if there was traffic in a "holding" pattern, or fog, smog or bad weather. Jo worked her turn at the Evansville, Indiana, factory also.

Jo reflects that she would have flown airplanes for no pay at all. The P-47 was undoubtedly her favorite pursuit plane. She regrets she never got a chance to fly the P-38 Lockheed Lightning. She regrets that WASP were dismissed before the P-47N the latest model was on the line, and she did not get to fly it. She did fly the old models, some with the olive drab paint. She recalls flying British lend-lease P-47s from the factory. The British planes had a bullseye painted on the side.

✭✭✭

Celia Hunter remembers Jo Pitz made a forced landing at Newark on one of the shuttle runs between Farmingdale and Newark — her P-47 had a piece get loose in the supercharger and she was going lower and lower as she lost power, but she was cleared for a straight-in approach and got it down ok. "None of us could track her in the fog; we just heard her and the tower talking."

Celia M. Hunter

Nancy Lee Baker, Jill McCormick and Virgie Lee Jowell, of the 2nd Ferrying Group, Virginia Clair and Grace Birge of the 3rd Ferrying Group on the wing of a Republic P-47. (Virginia Lee Jowell collection)

Virginia "Tex" Clair dons WASP typical winter flying gear with parachute to deliver a P-47 Republic Thunderbolt. Though based at the 3rd Ferrying Group, she was TDY to the 2nd. The gloves are her own since gloves were never issued. (Virginia Clair collection)

Virginia Lee Jowell

Virginia Lee Jowell recalls that in August 1944 a group of 13 WASP were sent from 2nd Ferrying Group, Wilmington, Delaware, to Evansville, Indiana, to ferry the Thunderbolt P-47s. She was first detached from the 3rd Ferrying Group at Romulus, Michigan and sent to Wilmington so that she could join the group of women pilots since her "new" husband had recently been assigned to Evansville to test the P-47s. They married on July 25, 1944 at Romulus AFB.

"What fun! Our usual run, a milk-run, we called it, was to deliver a P-47 to Newark in one day, catch Eastern Airlines out of LaGuardia that same night, and arrive back in Evansville by 3:00 a.m. the next morning. So, in addition to Teresa James and her husband, my husband and I all flew P-47s. The 13 WASP shared a big house, and I had an apartment with my husband.

When the WASP were disbanded in December 1944, the group flew back to Wilmington for discharge. I have no papers from those days, just a few photographs. I do have some beautiful memories."

Nancy Lee Baker

Nancy Lee Baker, WASP class 43-W-4, Wilmington, tells about the shuttle of P-47s from the factory to the port of embarkation.

We would fly from Wilmington over to Farmingdale, the factory in Long Island, to do the shuttle. Teresa James and Gerti Meserve would take us over in the Lockheed. They acted like they were fighting, and we wondered who in the devil was flying the airplane. I didn't find out until 20 years later that this was all put on. After we arrived at Farmingdale, we'd all sit until the weather got good. With all the fog and having to fly over New York City, you couldn't get off the ground. I remember I didn't like the wait ...the wait was bad. You'd wait for the weather; and when it did clear, you really had to get off the ground fast.

I do remember one of the funny things: there were only so many seats and not everybody could sit down. One of Helen Richey's things was that when everybody had a seat and she didn't, she had a glass of water she'd tinkle back and forth. This was to make people have to go to the "john" through the power of suggestion; and it worked, I guess.

One of the dangers Wilmington had was taking off and flying and landing within a half hour or so, and sometimes you had to circle-circle-circle. Sometimes as you were circling because so many airplanes were coming in, the fog might also be coming in and you had so much to look after with all these pursuits flying around Newark Airport to get in. I don't remember any accidents of any of our Wilmington group of women or of any of the fellows or anybody else.

Nancy Lee Baker wears the official WASP uniform. Note the diamond shape centered between wings. (Nancy Lee Baker collection)

We only flew from Farmingdale to Long Island and New Jersey. The P-47 shuttle was a very short distance, and usually when the weather cleared ...good Lord... as a matter of fact, there was a record there one day. I knew the guy who worked in the tower. When the weather cleared, can you imagine, craft all over the area waiting to get into Newark. When the weather cleared, **everybody** was up in the air at one time. That's kind of treacherous. And then the weather might be marginal, the visibility over New York — the haze and everything — so it wasn't the most delightful. I guess none of us ever said that when we were scared or that we didn't like it.

I know I didn't like the waiting. I'd get an nasty tummy from the waiting. And when the weather did break, and as I said quite often it was marginal, there were all those airplanes in the air, plus all those people all over the area waiting to get into Newark. In Newark they would "pickle" the airplanes...take the wings off and get ready to send them by ship overseas.

Women in Pursuit

Screen actress Irene Manning is visited by her cousin, WASP Nancy Lee Baker, on the Warner Bros. Studio's set of "The Doughgirls." (Nancy Lee Baker collection)

Cartoon sketch by A.J. May from her book RON.

Nancy Lee Baker waves from the cockpit of a P-47 Republic Thunderbolt, typical of those she ferried regularly. (Nancy Lee Baker collection)

But before I went into pursuit school and was subsequently sent to Wilmington, I went to Hagerstown, Maryland, picked up PT-19s and flew them all over the place, wherever they were needed. I went to Orlando for officers training. I had instrument training in Bryan, Texas.

I was at Houston. We started there. There were 43-W-1, -2, -3, -4 classes who started training at Houston in 1943. I think 43-W-4 was the first that started at Sweetwater. It was a double class, the largest class in fact. It had two sections: Flight A and Flight B. It seems that rivalry is just something that's just innate in people, and we had it between these two flights.

Betty Gillies, of course, sent us, and told us who was going where. She must have had faith in me. As I look back, earlier I was given an AT-6 to fly solo to California. I'd never been to California. You know people travel a lot **now** but they didn't used to. I ground-looped the damn thing somewhere in the desert at a refueling stop. I felt so bad about it. Also you have to take these tests to see if you'd had anything to drink. It was in place with lots of WASP so it may have been in Dallas. Anyhow, I got back to Wilmington and had to report to the Safety Officer. He asked what kind of a ground loop I did and I said, "A good one!"

I thought, well, I'll never get to California. But Betty gave me another AT-6 the next week to take out to California, and I was a lot more careful. I was perhaps careful to begin with, but I was a lot more appreciative of the fact.

Apparently Betty was the one who sent me down to pursuit school. I was probably a little scared, maybe all of us were a little bit, though I never would have said I was. You have to admire the fellows, though, who did say they were scared. Some fellows refused and that takes a lot of guts.

I thought I don't need to worry anyhow because I wouldn't pass. So I was relaxed. I had an instructor named Stanfield at pursuit school. I didn't like him at all. He was terrible. He screamed all the time. They put us in the back seat of a BC-1 or AT-6 to fly, and he would scream and carry on. I think I took the headphones off half the time. Well, I knew I wasn't going to get through this thing, so I was relaxed as you could be. I remember one of the guys who flew with the same instructor after me. He said that they didn't like to come after he'd flown with me, because he was so mad.

I just sort of vaguely remember this, and I was sure I wouldn't pass. You know how they worked that ... you come down to the flight board in the morning and see your name up there to fly such and such, and all of a sudden you come down one day and see you're scheduled to go on a P-47. As I recall, we had gone through all the processes, including the blindfold cockpit check, and we had had the ground school. I guess the reason I got through was that I was so damn sure I wasn't going to get through that I was so relaxed.

It was kind of a shocker to come down and see your name posted that you're going to go out in one of these damn fool things by yourself. We took them up and had to stay up at least an hour, if not two hours. We had to do all those stalls, which was good. Kids don't have to do all this in airplanes now, and it's terrible they have accidents today in these small planes. We had to put the fighters though all these steps so that when you came down to land you had a good feel for the airplane.

One other thing I remember about pursuit school is that I did get called into the Safety Officer once. They watch you land and take notes. I imagine that like many others I had to go around the field while they talked me in, like I might be too "hot" or something. But one thing I do remember: I bounced a P-40 and I had to recover with gunning it and using the right rudder. I thought nothing of it. I made a

Taking time to pose for a photo while at Republic P-47 factory are (standing) Teresa James, unidentified Republic factory representative, Liz Pearce, Gertrude Meserve, a second factory rep. and Pat Lawler. Seated are Anna Flynn, Ginny Alleman, Rita Moynahan and Janet Zuckowski.

good recovery. The Safety Officer called me over and apparently I'd done something that a lot of people don't live through. I didn't know that. They don't live through recovering a P-40 because the torque was supposed to be so bad that you can't do it. That's just one of the things I remember, that my recovery was okay.

I was thrilled upon graduating from pursuit school finally after thinking that I wouldn't do it. It was probably a real surprise except maybe thinking that now I'm going to fly 'em. And I enjoyed flying 'em then, I will say that. The only thing I do remember is that at Farmingdale the bad part of the P-47 shuttle for us was having so much time to sit and wait, play cards or do something. And when they said "Go," you had to really get out there and go before the weather closed in again. To me I found waiting really nerve-wracking.

Having just learned that 96 women graduated from fighter school at Brownsville and 56 didn't, I just assumed at the time that all of us were doing pursuit ferrying. I didn't realize so few of us were doing it. So we are very unique women.

★★★

After the war, some of us, including Helen McGivery, one of the original WAFS, used to get together at Sis Bernheim's house. One time when we were in a bar batting the breeze, Helen and I got to talking. I recited a poem. Helen knew the poem, too.

Renascence
All I could see from where I stood
Was three long mountains and a wood;
I turned and looked another way,
And saw three islands in a bay.
So with my eyes I traced the line
Of the horizon, thin and fine,
Straight around til I was come
Back to where I'd started from;
And all I saw from where I stood
Was three long mountains and a wood.
Edna St. Vincent Millay
Collected Poems

Do you remember that poem? It is Renascence by Edna St. Vincent Millay. It's a long poem. That's one of the verses. And I guess the thing is how dead you... she was... and she came back... mentally... but it stuck out as a very deep, close point between McGilvery and I. She was killed soon after that. She was in a mid-air collision in Long Island (after the WASP deactivated). She was instructing, I believe. You can have a more intimate conversation under certain circumstances than you can have in others, and we both struck very deep. Anyway, that's what happened there, and I will always remember it.

★★★

During the WASP I looked up to the WAFS originals. I knew they knew so much more than I did; they had so much more air time. Helen McGilvery gave me hell once for being off course.

Ferrying was a wildish deal.

Helen Mary Clark

Helen Mary Clark, WAFS from Wilmington Base, 2nd Ferrying Group, went to Long Beach on TDY (temporary duty) just to get checked out in the P-38. She remembers her "hilarious" adventure of getting several P-38 trainers in a row, all of which had to have the gear pumped down (150 arm-aching pumps), and other emergency procedures.

Teresa James

History reports that by the end of November 1944, women pilots had delivered 1,987 P-47s. Teresa James, an early WAFS, had the honor of making it 2000 by delivering that numbered Thunderbolt off the assembly line.

Ruth Adams

Ruth Adams

"I have checked my logbook, and my first trip after graduating from Brownsville Pursuit School on August 22, 1944, was in a P-51C, the only plane I had **not** flown at Brownsville. (Some earlier pilot trainee had washed-out the trainer 51.) As I recall, Captain Cox went over the cockpit with me (cockpit check) for an hour at Hensley Field, Texas, the North American factory site. Then I took it up, made some turns and a stall or two at 10,000 feet, and brought it back so Captain Cox could watch my landing. He was upset that I was gone so long—one hour. Then I took the P-51C from Hensley to Love that day, a half-hour flight. The next day I proceeded from Love Field at Dallas to El Paso, the refueling stop, and on to the California North American modification factory."

"I was co-piloting for a WAFS, Esther Rathfelder Manning, my first order upon being sent to New Castle Army Air

Petite Ruth Adams, wearing her flight suit, leaves the P-51 she has just delivered, carrying her parachute over her shoulders, and the planes clearance and delivery papers in her hand to the operations office. (Ruth Adams collection)

Base after graduation from Avenger Field. I think they wanted to get an impression of whether this WASP could fly before sending me out on my own. Anyway, we had just taken off, and were no more than 100 ft. up when I noticed smoke coming up from under the pilot's seat. I told her we had a fire, but I thought we could make it back to the field (as I saw no open flame.) The fire extinguisher was on the wall behind her seat. I opened the door to see how much altitude we had and whether we were high enough to jump if we had to. I saw then that we had only about 100 feet, clearly not sufficient altitude to allow the chutes to open.

Although the fire extinguisher was supposed to be operated at a 30 foot distance from any fire (the instructions said), I had no choice, and proceeded to use it from five feet, to keep the fire from spreading. Then I saw that we were going into a field straight ahead; and being out of my seat without time to strap myself down, I braced my back, with my hands behind my head, against the seat like I had seen diagrammed for emergency procedures. Esther made a neat landing on some farmer's flat field. We didn't even go up on our nose.

We heard later that we had had an electrical fire. Probably the insulation on wiring to the landing gear was burning, and

The 2nd Ferrying Group WASP gather to bid each other and their "room mother" a fond farewell, the day prior to official deactivation of the WASP on December 20, 1944. (Betty Scantland collection)

The last supper of the 2nd Ferrying Group WASP as a group at Wilmington the day prior to official deactivation of the WASP on December 20, 1944. Out of photo range on left were Jane Straughan and Marjorie Gray; standee, unIDentified; Gert Meserve, Helen McGilvery; standee, unID; Nancy Batson, Nancy Love; unID; Betty Gillies, Helen Mary Clark, Teresa James, Nancy Lee Baker; standee and 2 next seated unID; Emily Hiester; seated unID; Gwen Cowart.
(Nancy Lee Baker collection)

it probably started when we tried to raise the wheels right after take-off. The fire stopped of its own accord (maybe when she turned off the ignition). The fire did not seem life-threatening to me, even though we were in the "bamboo bomber," because I never saw open flame. The surgeon's office report was correct. I was not upset. Just disappointed that we didn't complete the delivery of the UC-78. After Esther and I landed, I volunteered to stay by the plane to "guard it" while she went for help. I understand that when she wrote up the accident later, she had good things to report about my conduct. My height: 63-1/4 inches. A couple of incidents in flying pursuit after Brownsville scared me, but that's another story.

✭✭✭

I don't know which WASP crashed into the bay. It is news to me that anyone did.

Nancy Batson

The following article reprint, from the collection of Nancy Batson, appeared in the November 1987 WW 2 Times. The original date and publication information was unknown at the time this book was printed.

Wasp Wins Struggle, Lands Plane Safely

—Post-Gazette Photo
Wasp Nancy Batson shows Captain Jack Clay her hands. She blistered them badly, but saved her P-38 plane.

Fights Faulty Gear While Circling Airport, Radios for Crash Truck, Avoids Accident

After struggling with defective landing gear for more than two hours while she circled county airport, Wasp Nancy Batson, 24, grimly radioed the word to "get the crash truck out" and then nosed her P-38 pursuit plane down to the landing field.

While airport attaches watched breathlessly, Miss Batson hit the runway with her fuselage landing wheels and kept the nose of the plane high, fearful that the nose wheel had not locked properly.

Gingerly she settled—and the wheel didn't fold up. After the plane taxied to a stop, she climbed out of the cockpit and said simply, "I'm sorry to have caused all this trouble."

Captain Jack Clay of the Air Transport Command, who had instructed the plucky Wasp by radio how to adjust the faulty gear, grinned. "You're calmer than a lot of men this has happened to," he said.

Miss Batson was ferrying the plane from Pittsburgh to Newark when she noticed that the hydraulic pressure, which operates the landing gear, was down. She headed back to the airport.

From the control tower, Captain Clay told her what to do. After circling the field several times, she pumped the pressure by hand to the point where the fuselage wheels were lowered. Then the nose wheel went down.

But neither she nor anyone below could tell whether it was rigidly locked.

After it was all over, Miss Batson did the thing any woman—not necessarily a Wasp—would do. With hands blistered from pumping the landing gear, she proceeded to freshen her lipstick.

Ruth E. Anderson

Gwendolyne E. Cowart

Virginia Clair

Irene G. Gregory

Rosalie L. Grohman

Emily Hiester

Mitchell Long

Marian G. Mann

Jill McCormick

Rita Joan Moynahan

Sarah Elizabeth Pearce

Mary M. Rosso

Janet J. Zuchowski

3rd Ferrying Group

3rd Ferrrying Group
Air Transport Command, Romulus, Michigan

Name	WASP Class	Pursuit School	Graduation Date
Archibald, Betty J.	43-W-3	44-9B	5/15/44
Barrick, Helen F.	43-W-7	44-18B	10/1/44
Birge, Grace	43-W-3	44-10B	6/1/44
Bolish, Juanita	43-W-4	44-9B	5/15/44
Bowden, Mary Louise	43-W-4	44-19B	10/15/44
Callaghan, Claire G.	43-W-1	44-8A	5/1/44
Conner, Lillian M.	43-W-3	44-10B & 44-14A	8/1/44
Crinklaw, Virginia B.	43-W-3	44-17A	9/15/44
Dahmes, Sylvia	43-W-5	44-18B	10/1/44
Darling, Mary	43-W-2	44-8B	5/1/44
Dickerson, Patricia A. "Pat"	43-W-2	44-12A	6/30/44
Donahue, Barbara L. "Donnie" (Squadron Leader)	WAFS	44-4	2/21/44
Gery, Ellen H.	43-W-2	44-8A	5/1/44
Hamilton, Margaret A.	43-W-2	44-8B	5/1/44
Johnson, Mary Catharine "Jary"	43-W-2	44-4	2/21/44
Kerr, Margaret E.	43-W-2	44-8B	5/1/44
Ketcham, Marjorie	43-W-1	44-5	3/8/44
Lawler, Florence P. "Pat"	43-W-2	44-8B	5/1/44
Lee, Hazel Ying "Ah Ying"	43-W-4	44-18B	10/1/44
Lovejoy, Alice	43-W-5	44-12A	6/30/44
May, Alice Jean "A.J."	43-W-4	44-12B	6/30/44
Miller, Maurine "Mimi"	44-W-4	44-18B	10/1/44
Myers, Marcie Jo "Jo"	43-W-5	44-13A	7/15/44
Scantland, Mary E. "Betty"	43-W-6	44-13B	7/15/44
Scott, Jane S.	43-W-5	44-18B	10/1/44
Shea, Elizabeth "Betty"	43-W-5	44-9A	5/15/44
Snyder, Frances M. "Fran"	43-W-6	44-18B	10/1/44
Sweet, Virginia L.	43-W-4	44-19A	10/15/44
Trebtoske, Joan M.	43-W-4	44-18B	10/1/44

Barbara L. Donahue

Joan M. Trebtoske

Barbara Donahue

Barbara Donahue served as Squadron Leader at Romulus after Adela Scharr. She was in pursuit school class with Emily Hiester and Jary Johnson.

Juanita Bolish recalls Barbara Donahue as a marvelous person and a good leader of the WASP Romulus Squadron. "Barbara wore much makeup." Another WASP described Barbara as "most glamorous."

Joan Trebtoske

When Joan graduated from the WASP class 43-W-4, she was assigned to the 3rd Ferrying Division of the Air Transport Command at Romulus, Michigan, near Detroit. After ferrying training airplanes for a year, she was upgraded in larger horsepower craft, obtained her instrument rating, and was sent to Brownsville, Texas, in October 1944 to learn to fly the high-speed fighters All seven WASP in her class were from Romulus. She checked out in the P-39, P-40, P-51 and P-63 at Brownsville Pursuit School. Joan's favorite airplane was the P-51 Mustang because, she said, it was sleek to fly and fast.

But, she said, flying was not what most people imagined. "Many people thought it was so glamorous, but it really wasn't. It was also hard work. There was

Joan Trebtoske boards her plane at Romulus. (Joan Trebtoske collection)

always a risk. You never knew when the engine would give in; but I was not really worried unless I heard the engine miss."

While at pursuit school, she had her first forced-landing in a P-47. "The master cylinder blew and I had no power. That wasn't easy," Joan recalled.

She remembers the freedom of the skies with a tinge of nostalgia: "There are days when the sky is a certain way, and I think: this is how it was when I flew this aircraft — some nostalgia is there, once in a while."

Jary Johnson

"I was stationed at Romulus, Michigan when the training at Houston was finished. After flying liaison and other light planes for a few months, I

Mary Catharine "Jary" Johnson

Jary Johnson stands beside a P-40 Curtiss Kittyhawk, a plane she frequently ferried (Jary Johnson collection)

earned an instrument rating and went to Palm Springs for pursuit training."

"For the rest of the WASP service, I ferried P-40s, P-51s, P-47s, P-39s and P-63s. Because of Romulus' proximity to Niagara Falls and Buffalo, most of my deliveries were P-39s and P-63s from the Bell factory and P-40s from the Curtiss factory."

"Dottie Nichols, who was my friend in Van Nuys, California, before the war, was responsible for my joining the WASP. Because I had only 60 hours when Jacqueline Cochran started the training course and Dottie had only 30, we quit our jobs and drove to Fort Worth to accumulate flying time as fast as possible. (Flying had been shut down in the coastal section of California at the time of Pearl Harbor.) For a reason that was never revealed to us, we were accepted as trainees only two weeks later, with far less than the required 200 hours flying time."

Dottie and I roomed together in Houston during training; and then because she was stationed in Long Beach, we didn't see each other for some time afterward. Our first and only reunion was when we happened to be delivering P-39s to Great Falls, Montana, at the same time. We were delayed for two weeks in Bismarck, North Dakota, along with a number of other pilots from various bases, because of weather, and we shared a room in the hotel during that time.

One morning, we went to church together and then were notified that the weather would allow us to continue to Great Falls. She was ahead of me in the line of P-39s waiting to take off. When she was about 40 feet in the air, her engine quit and her plane crashed and burned on the runway. Because I had been her friend for along time, I was assigned the sad task of accompanying her body home to Los Angeles by train, and delivering it to her mother. She was given a military-style funeral in Van Nuys, with five other uniformed WASP besides myself as pallbearers; and she was buried in Forest Lawn Cemetery.

After I released my plane in Bismarck, it was assigned to a male pilot who was flown in from Romulus. I learned later that the engine quit at 500 feet after he took off, and that he was high enough to land in a field and survive. The rumor went around that sabotage had been responsible, because they found water in the gas; but, of course, we never knew for sure.

Margaret Ann Hamilton's eyes twinkle as she flashes a winning WASP smile. (Margaret Ann Hamilton collection)

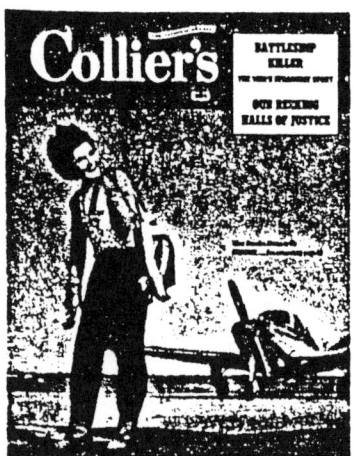

Collier's for April 2, 1949

This week's cover: Photographer David Peskin found Ann Hamilton, the girl on the cover, at Roosevelt Field, Long Island, New York. As she strode away from her plane, he saw in the combination of space, strength and freedom, a suggestion of the spirit of America. And when he brought the picture to Collier's, we saw the same quality. We asked Miss Hamilton what America meant to her. This is what she told us: "Say 'America' to me and I see a great sweep of wonderful color that seems to fill the world. I was born in Enid, Oklahoma. Out there, everywhere a child turned, he faced the United States. For me there was never any end to America. There still isn't.

"I'm a flier. As a Wasp during the war, I flew just about everything they'd let a woman handle. I know what it is to sit in the sky feeling the thrill of power in your hands. By the time you read this, I'll be in Japan with the military government helping to get the idea of democracy across to people who were once our enemies.

"For America, this job is something like sitting in the sky with tremendous power at your command. The U.S. could take advantage of its strength, but it doesn't—for all of us have grown up with respect for the liberties of groups and individuals. If there are going to be storms ahead, it will be our duty to stay on the beam, to keep our sights on basic human rights. If we do that, we'll continue to be what I call the best bunch of people under God's blue sky."

Ellen Gery stands beneath the air intake of a P-40 nose. (Claire Callahan collection)

Mary Darling pauses from reading her map to smile for the camera. (Lee Huber (Mary's son) collection)

Mary Darling and Betty Whitlow with parachute, flight bag, headsets and maps by a twin-engine Cessna Bobcat. (Betty Whitlow collection)

Betty Shea

Betty was based in Romulus with the 3rd Ferrying Group and graduated from the Brownsville, Texas Pursuit School on May 15, 1944.

Dottie Nichols, classmate in that same pursuit group of 14 women pilots in two flights, graduated on that same day; and was based with the 6th Ferrying Group at Long Beach. Both Betty and Dottie were in Bismarck with the many pilots who were moving Bell Aircobra P-39s to Great Falls, Montana, from the Niagara Falls, New York factory.

On June 11, Dottie was killed in a P-39 on take off at Bismarck, North Dakota, when her engine quit. Jary Johnson accompanied her body back to her family in California. Betty recalls "We were in a bunch of troubled airplanes at that time."

Pilots were piling into the Bismarck Prince Hotel because the weather was so bad. "We were held for the investigation of the crash. The only nice thing about the memories of the Prince Hotel is that they would wash and iron a shirt overnight."

Romulus Squadron Leader, Barbara Donahue was there, as was Jary Johnson, Juanita Bolish and about ten WASP — all with Bell airplanes, and all going to Great Falls, Montana.

Romulus pursuit pilots flew in groups; Betty recalls that her flight leader was an officer named Hanna.

The investigation of Dorothy Nichols' engine failure on take off finally concluded. The cause may never be made public, but sabotage was suspected.

The weather cleared, and pilots left for their destination: Great Falls, Montana.

★★★

Betty left Bismarck, and only 20 minutes out of Great Falls **her** engine quit. It was just outside Hobson, Montana, on June 19, just eight days after Dorothy Nichols' death, Betty Shea was forced to bail out of her P-39.

Betty Shea poses for WASP photo wearing the official uniform complete with insignia on hat, buttons on shoulders, WASP and propellor on each lapel, and the diamond-shape-between-wings emblem above the pocket flap. (Betty Shea collection)

Fran Snyder, wears an Eisenhower type battle jacket, part of the WASP uniform. (Fran Snyder collection)

Juanita Bolish

Juanita Bolish, WASP class 43-W-4 Pursuit Pilot, Romulus, recalls:

"I wasn't interested in having a date; we'd meet somewhere, maybe eat together, have a drink together. I drank very little. It was so wonderful doing the job we were doing, so exciting, that I wouldn't have traded places with Eleanor Roosevelt or anybody in the United States. That's the way it was. And almost every woman pilot felt the same.

"We'd get together. I never did play cards, or anything like that the way some of them did at Romulus Base, I didn't bowl."

"One time I was going to Canada. We flew planes for Canada, from Canada, around Canada, and to Canada. (Romulus was near Detroit and the closest ferrying base to Canada.) I think I was in Minneapolis or St. Paul when I went to watch them ice skate. It was very cold there. It's strange, the little things that are so memorable."

Once when Juanita was stuck in Bismarck with the weather, she read *The Sun is My Undoing*. She also visited nearby historic Mandan. She took high heels in the cartridge case in the pursuit wing and a skirt in her briefcase; she never packed a gun.

One trip she flew a P-63 as wingman and remembers landing in Yakima, Washington, on the way to make a delivery in Portland, Oregon. She remembers always being afraid and having no self-confidence.

Betty Archibald in cockpit of a P-63 reviews her flight map with Juanita Bolish on wing. This particular plane is unpainted, then waxed to make it faster. (Smithsonian collection)

Claire Callaghan checks the instrument panel in her plane prior to take-off. (Claire Callaghan collection)

Sylvia A. Dahmes

Virginia L. Sweet

Jane S. Scott

Betty Archibald, Patricia Dickerson, Janet Zuchowski and Alice Jean May and a P-63A Bell King cobra on a bright, sunny day. (Alice May collection).

Helen Barrick pauses for a photo in her plane. (Helen Barrick collection)

Sylvia Dahmes, WASP photo in full uniform. (Sylvia Dahmes collection)

Ferry pilots by a P-63 Kingcobra at the Bell Factory, Niagara Falls, NY. Betty Archibald, Juanita Bolish, Rena Wilkes and Janet Zuchowski with 2 unidentified men.

Maurine Miller

Maurine "Mimi" Miller got her private license through the CPT program at Northwestern University, Evanston, Illinois. She went on to get her commercial and instructor's rating, and was instructing at Palwandee Airport north of Chicago when the war began. Having been told by her doctor that she had a heart murmur, she did not join the WASP program when it began. She was called to instruct on primary trainers at Sweetwater, Texas.

"Mrs. Deaton told me to join. I did not have a heart murmur." She joined WASP training class 44-W-6, was boosted up to 44-W-4 class, which graduated May 23, 1944. Stationed at Romulus, 2nd Ferrying Group, she was sent to Brownsville for pursuit training and graduated pursuits September 15, 1944. Mimi spent three months ferrying fighters.

Helen Barrick wrote that she and Mimi roomed together on two P-63 trips—one was the trip when Hazel Ying Lee was killed.

Maurine "Mimi" Miller with flight instructor "Smitty" beside the Sweetwater transportion bus. (Doris Tanner collection)

Maurine "Mimi" Miller

Louise Bowden has just delivered a radio-controlled, plywood-built PQ-8 to Chanute Field, Kansas, Spring of 1944, and explains her ferrying job to the WAC Officer stationed there. (Louise Bowden collection)

Louise Bowden

Louise Bowden graduated from pursuit school on October 15, 1944. She relates the following.

"I bailed out of a P-51, not the P-39. It was December 10, 1944. The engine stopped and started several times before it stopped for good. I remember the last time I saw the altimeter it read 1500 feet. I never had time to look up and admire the parachute. I remember calling the tower and saying I was leaving the plane. The hatch knocked my head set off ... this is reversed ... I pulled the hatch handle; and when my headset was gone, I told the tower I was getting out. I remember seeing my feet in the air and beyond my feet a two-engine Navy plane. I felt they saw me."

"I was momentarily knocked out when the chute opened. Then I had to steer the chute away from the burning plane. I landed about 40 yards from it in a harvested

Young Aviatrix Bails Out of Burning Plane

LOUISE BOWDEN

Louise Bowden, of Elmer, young aviatrix, had an exciting experience while with the WASPS when she parachuted to safety from a burning Army fighter plane near Greensboro, N. C. It was the pilot's first forced jump in more than three years of flying.

According to an Army press relations release, witnesses reported that the plane was in difficulty for a time prior to the crash. Finally Miss Bowden notified the Greensboro-High Point airport that she was in trouble and was bailing out. It was reported that the ship was afire before the pilot bailed out but this was not verified by the Army.

The plane crashed in a plowed field. The single motor was buried in the ground and a minor explosion followed, turning the craft into a roaring inferno within a few minutes. The pilot, who bailed out at about 1500 feet and landed about 20 yards from the burning ship, was uninjured.

Miss Bowden had qualified for a civilian pilot's license before going to Sweetwater, Texas, to fly with the WASPS. She received her training at Buck's Airfield, at Woodruff, with everyone around the field glad to help the little redheaded girl who was so eager to learn to fly. She was a technician at Bridgeton Hospital before making flying her career.

The young woman pilot has flown every type of Army plane including the speedy fighters which require great skill to handle.

corn field near High Point Army Air Force Base, North Carolina. After landing, I looked up and saw the twin-engine Navy plane circling. I wanted to say, "I'm okay!"

They called the Army field I had left. Then I saw an Army Lieutenant running toward me. He had pointed out the Navy plane to his wife as they drove by, and as she looked up she saw me coming down.

They took me to an Army hospital emergency room where I had treatment for delayed shock. Besides that, I only scratched an ankle. I should say all the emergency procedures that the Army required us to memorize came to mind when I needed them. They gave us written tests on emergency procedures for the five fighters we were checking out in pursuit school at Brownsville, Texas.

I did not see the report of this accident and did not learn the cause of engine failure. This was my only P-51 Mustang delivery. I had P-39s, P-63s and P-47s to deliver.

After the accident date of December 10, 1944, there were only ten days left before the WASP were deactivated and women's flying skills no longer needed. I delivered small planes until the WASP were terminated.

M. Louise Bowden

Hazel Y. Lee

Hazel Ying Lee

Hazel "Ah Ying" Lee was always laughing, outgoing, enthusiastic. She was slender, "boyish," maybe 5'5," had protruding teeth — she was much fun. One time she had a forced landing. The farmer thought she was a Japanese invader and approached her with a pitchfork. She got excited and yelled to the farmer, "I China Girl, I China Girl."

★★★

(Author's note: Hazel was killed in a mid-air collision at Great Falls, Montana, on November 23, 1944. I am currently compiling the information surrounding this event which will constitute a separate book.)

Women in Pursuit

Cartoon sketch by
A.J. May from her book RON.

Helen Barrick

Jo Myers

Alice J. May

Alice E. Lovejoy

Alice Lovejoy

Alice Lovejoy was an employee at the Piper Aircraft plant at Lockhaven, Pennsylvania in 1942. Prior to joining the WASP, she was instructed in flying by Lauretta Beaty, who had also taught Nancy Baker to fly at the same Piper plant. Eighteen Piper women joined the WASP and, of those, five went on to Pursuit School: Alice, Lauretta, Nancy, Anna Flynn and Lola Perkins.

After six months training at Avenger Field near the wind-swept town of Sweetwater, Texas, Alice was one of 85 to graduate September 11, 1943, class of 43-W-5.

She was assigned to the 3rd Ferrying Group at Romulus, Michigan. After serving a year, she was sent to pursuit school at Brownsville, Texas; and there she perished in a mid-air collision in an AT-6 on September 13, 1944.

Louise Bowden and Jill McCormick were classmates in that same pursuit group. In a later interview, Louise stated the AT-6s were flying in formation; she believes a wing tip of the next plane hit her, and the instructor with Alice bailed out. He suffered afterward, believing that Alice's hatch jammed.

Louise accompanied Alice's body from Brownsville, Texas, to Scarsdale, New York; and at every train stop, she relates that she went back to the baggage car to make certain the body box was still there.

Louise tells that Alice was well-liked, upbeat and personable—a nice person. She was well-groomed, a refined person, a little bit on the plump side. Louise met her folks. Alice's father was tall and distinguished. They lived in a lovely, old-fashioned home. Louise wore her uniform at the service, and remembers walking down the church aisle with a lieutenant, also in uniform.

NORTH AMERICAN TRAINER AT 6

Women in Pursuit

Grace Pitkin Birge

Claire G. Callaghan

Virginia B. Crinklaw

Lillian Conner

Patricia A. Dickerson

Ellen H. Gery

Margaret E. Kerr — Marjorie Jane Ketcham — Pat Lawler

Romulus last day group photo. Dec. 1944 AJ. May, Lil Conner, Fran Snyder, Juanita Bolish, VA. Malany; seated 2nd row, Pat Dickerson, Betty Shea, Lenore McElroy, Barbara Donahue, Margaret Ann Hamilton, Joan Trebtoske; seated 3rd row, Bert Miller, Ellen Gery, Claire Callaghan, Louise Bowden, Betty Archibald, Sylvia Dahmes, Virginia Sweet, Helen Barrick, Virginia Crinklaw; standing rear "Mimi" Miller, Jane Scott, Jary Johnson, Margaret Kerr, Pat Chadwick, Grace Clark, Margaret McCormick. (Betty Scantland collection)

WASP 3rd Ferrying Group at Romulus, Michigan, December 1944 farewell banquet. Squadron leader Barbara Donahue(center) looks up from buffet table while other WASP mingle in the background. Clockwise Lenore McElroy, Virginia Sweet, Pat Chadwick, Pat Dickerson, unID'd male, 4 unID'd WASP at back of room, Margaret Ann Hamilton and Juanita Bolish. (Sylvia Dahmes collection)

Dorothy Scott, WAFS, killed in a mid-air collision at Pursuit School, Palm Springs.

5th
Ferrying Group

5th Ferrrying Group
Air Transport Command, Love Field, Dallas, Texas

Name	WASP Class	Pursuit School	Graduation Date
Abell, Helen T.	43-W-6	44-17B	10/1/44
Bachman, Betty Jane "BJ"	43-W-2	44-2	1/44
Bevis, Gayle D.	43-W-5	44-12A	5/15/44
Bohn, Delphine (Squadron Leader)	WAFS	—	
Carlstrom, Marion	43-W-5	44-10A	6/1/44
Carmody, Nelle L.	43-W-6	44-17B	10/1/44
Colby, Cornelia Y. "Connie"	43-W-4	44-8B	5/1/44
Dailey, Ruth	43-W-2	44-5	3/8/44
Emerson, Norma Jane "Jane"	43-W-2	44-3	1/44
Fremd, Helen Irene	43-W-5	44-10A	6/1/44
Fullwood, Rosa Lea	43-W-4	44-11B	6/16/44
Gorman, Gretchen J.	43-W-3	44-8B	5/1/44
Gott, Lois Kathrin "Kay"	43-W-2	44-5	3/8/44
Hopkins, Dorothy	43-W-6	44-17A	9/15/44
Kary, Ross "Rossie"	43-W-2	44-5	3/8/44
Lindley, Ruth	43-W-5	44-12A	6/30/44
Luttrell, Virginia D. "Ginny"	43-W-4	44-10A	6/1/44
Madison, Isabel "Pete"	43-W-4	44-9A	5/15/44
Miller, Sidney	43-W-1	44-2	1/44
Schaefer, Helen M.	43-W-4	44-16B	8/31/44
Scott, Dorothy	WAFS	44-2	
Tate, Janice R. "Jan"	43-W-4	44-9B	5/15/44
Thurn, Violet C. "Vi"	43-W-4	44-9A	5/15/44
Tomlinson, Evelyn L.	43-W-6	44-9B	5/15/44
Tompkins, Gertrude V. "Tommy"	43-W-7	44-17B	10/1/44
Turner, Helen A.	43-W-5	44-12B	6/30/44
Whitlow, Elizabeth "Betty" or "Bee Whit"	43-W-2	44-3	1/44
Willis, Barbara	43-W-4	44-8B	5/1/44

Delphine Bohn, WASP squadron leader.

Ruth Dailey

The following are recollections in an interview on February 16, 1987.

Ruth: I was co-pilot for Delphine Bohn on B-17s. I was called in to Captain Garrett for my choice to fly heavy multi-engine ships or pursuits. He tried to persuade me into multi-engine, since he was an old airline pilot. I told him I wanted to try pursuits just as soon as I could. He said, "Why?" I said, "If something goes wrong, I don't want to wait for someone else to get out of my way, I'm going to get out!"

In pursuit school, we got separated in class and put in classes with all boys. We never saw another woman pilot except in the evenings.

I was there when Hazel Lee was killed. There must have been 50 planes stuck at Fargo. Everybody was there. It was snowing and very bad weather.

Kay: In those days, we were so individual. We were so busy we didn't keep track of who else was flying pursuits. Unless you were on orders with someone else, how would you know?

R: You wouldn't know.

K: Helen Schaefer was in pursuits, and she was one of the ones stuck in Fargo.

R: But she didn't fly P-38s.

K: As I recall the only reason we had P-38s at Dallas was the firing we heard from next door at Lockheed? Weren't they working with the guns?

R: I don't know.

K: I lived in the barracks furthest from the office, in the corner next to the Lockheed fence, so I always heard the firing at night when I was back on base. I just assumed the only reason the Lockheed Lightning had any goings-on there was to modify the P-38.

R: They did have modification there. They came from Long Beach, the West Coast. Do you know why they wouldn't let the girls fly them for so long?

K: No, why?

R: You don't know that they were losing the P-38s?

K: Oh, yes. I didn't know if I wanted to fly one or not. Jack Frost from Caldwell, Idaho, was killed in one — also Evelyn Sharp, one of the WAFS. We heard that

Ruth Dailey

Ruth Daily enters the cockpit of a P-38 as she prepares to fly it to its destination. (Ruth Daily collection)

Jack's plane had exploded in the air around Dayton, Ohio.

R: Do you know the whole story?

K: No.

R: They were getting them to Dallas successfully. When they left Dallas, they were crashing. The guys over at P-38 transition in Dallas put up a sign "Murder, Incorporated." They stuck it up because they were literally losing a lot of pilots. One of the experienced men pilots took a P-38 and he began to lose oil pressure; but he was experienced enough that he got the plane down somewhere. What they found was that somebody had put these minute holes in the oil lines, and had waxed over them.

K: Like sabotage?

R: Apparently; but when he got it down and they had a chance to check it, you see, they all had burned. Then one day CIC (Counter Intelligence) walked into the Lockheed Dallas Modification Plant, they walked up to this guy, picked him up, took him out, and nobody knows what happened.

Another trip I had a sabotaged plane, a P-38. Sidney Miller and I left Dallas together. We stopped in Amarillo, Texas just to top-off the tanks. You know how you do. I said "Sidney, let's go on to Denver, Colorado and spend the night" so I took off. I got to Denver and waited, and I waited and I waited, and she didn't show up. She lost an engine on take-off, and had to go back to Amarillo to land. So I sat in Denver.

That night in Denver it snowed. In the morning, I called and asked them to pre-flight the plane. They said "No need to get out of here before 10 or 11 since there is three feet of snow on the runway." Of course, I preferred not to fly alone. Sidney and I always tried to stay on trips together, and she and I both had ships for Cheyenne.

They pre-flighted my plane and had it ready to go by 10. I got out there at 10

Ruth Daily readies for take-off from the cockpit of a P-38. WASP delivered these planes from Long Beach factory to modification plants and to coastal defense stations as well as ports of embarkation for European and Pacific Theaters of war. (Ruth Daily collection)

o'clock and the head man on the flight line said I wasn't going anywhere. They found the sump frozen solid. I sat in Denver three days.

The plain clothes investigators said it was water in the gasoline. The CIC investigators told me they could determine three things: 1) If the water was merely condensation in the gas storage tank, it would all be distilled water; 2) if it wasn't distilled water, the water had been put in deliberately; then they could look for evidence of water in the gasoline where the tanks were topped-off in Denver, and 3) they could look for water in the gas at Amarillo, Texas.

Kay Gott

Kay remembers that her CPT instructor told her Mother, but never told her, that she had a good feel for flying and that she should continue in aviation. He said that many of his male students were too rough on the controls. After Kay earned her Private Pilot's license, CPT closed to women. He phoned Mrs. Gott to tell her that one of the ten owners of an airplane (a Porterfield with a 65 hp Lycoming engine) was going to war services shortly and would sell his share for half price — fifty dollars. Kay's Aunt Doris put up the money, and Kay went on to earn a Commercial Pilot's license in this airplane, flying Civil Air Patrol to build up the required 200 hours.

Kay's next step was working on a Flight Instructor Rating. This was nearly completed when she heard that Jacqueline Cochran wanted Commercial-licensed women pilots to fly for the government. She reported for training to fly "the Army way" in December 1942, and was in the second program admitted to Cochran's Women's Flying Training Detachment School at Houston, Texas.

Kay delivered training planes to schools for nine months before going to Pursuit School.

Kay Gott arrives at Houston, Texas, in December 1942 to learn to fly the "Army way."

She qualified in the P-38 on November 29, 1944, and had a total of 17 hours of P-38 flying time before deactivation of the WASP. In that time, she delivered three P-38s. Her last flight was delivering a P-38L from Long Beach to Dallas on December 13, 1944, picking up a North American F-6K at Dallas and delivering it that very same day to Meridian, Mississippi. VE Day (Victory in Europe) was 5 May 1945 — more than four months away. VJ Day (Victory in Japan) 14 August 1945 was eight months away!

✭✭✭

After nearly a year ferrying airplanes to training schools I was assigned to Instrument School at Love Field in Dallas. I worked on my "blind" flying ticket on the B-18 Bomber, the Douglas DC-3 and the Douglas C-47. The last two were both cargo work-horse airplanes with twin engines — very reliable, and easy for me to fly. Those Douglas airplanes are still flying today, some 50 years later.

In the Link instrument trainer, a box-like affair, we flew "make believe" in safety in the hangar. Nowadays this very sophisticated machinery is called a flight simulator; and the operator is able to control external things like wind, velocity, fuel

Kay poses for an official photo wearing Air Transport Command wings on her "Ike" jacket and Army Air Force gold eagle on her hat.

amount and air currents, as well as trace your flight path with a pen on the map. Thus you learn from any mistakes, and have a visual record of what you actually did in simulated flight.

For my final exam in the Link, I bracketed the beam, listening to the "A" and "N" station call in Morse Code, thereby I knew I was approaching the Salt Lake City

Kay Gott not in uniform this day, stands by a row of P-51Ds, her favorite plane to fly.

Douglas C-47 Cargo Skytrain wings its way to its destination.

airport from the North East and that I was high over the Uinta Mountains of Utah to let down over flat Salt Lake. Thus, I earned this part of my instrument card on an extremely difficult course.

The final exam in the air was a practical test flying a C-47 as first pilot. I was to deliver airplane parts: first a tire to Tulsa, Oklahoma, then on to another delivery in Oklahoma City, on to Fort Worth, Texas, and finally back to home base at Dallas. All of this while flying a C-47 under the hood with a blackout curtain hung over the window and all around me. This is roughly a rectangular course.

Flying "blind" is without visual reference to the ground. An observer sits in the righthand seat and watches the progress and the skill of the pilot flying the course, as well as accuracy following the route. A most important job for this observer is to keep an eye open for any traffic.

At home base, I did not have to land the airplane "blind." On the last leg coming into Dallas flying the beam, I had to come within 200 feet of the end of the runway. Now some years later, I wonder and marvel at just how I managed to do all this and with the equipment of 1944. But I passed both Link and practical tests and earned my Instrument Rating in February 1944.

I was confident in my ability to fly — no longer unsure about my competence.

I had just succeeded in difficult instrument school. Why should I fear failure in pursuit school? No way was I going to fail! I was a good pilot. A good ferrying pilot.

★★★

My first job after obtaining that Instrument Card was an assignment to co-pilot a C-47 twin-engine cargo airplane to Fort Wayne, Indiana. The pilot was "hung over," so we did not leave Dallas the day the orders were cut. Next day, we took off for Scott Field near St. Louis, Missouri, the first refueling landing. I was supposed to be checked out on the landing in a strange airfield. Well, the pilot flew the entire time, landed the airplane and instructed me not to touch the controls. On the next leg, the pilot again flew the airplane and again made the landing. By now, I was really protesting.

When we "sold" the airplane and inquired about transport back to home base Dallas we discovered we would have to RON.

The pilot drank all evening. I had one beer. He bewailed the fact that his "wife didn't understand" him. I got in the elevator to go up to my room and he shouldered his way into the lift. When we arrived at my floor, I pushed the "down" button and went to sit in the lobby. If I had to, I was prepared to spend the night in the lobby. I

P-51D Shimmy IV, flown by Col. Chet Sluder in Italy - he found Kay had delivered this plane from the factory to port of embarkation.

think the desk clerk was on my side. The pilot finally left, and I went up, alone, to my room.

When I got back to Dallas the next day, I stormed into operations where I knew one of the officers who assigned trips and was dating my WASP roommate. I told him never to send another WASP on a trip with that man.

✯✯✯

In March 1944, right out of pursuit school, I delivered three P-51s from Dallas to Newark and one P-40 Kittyhawk from the Buffalo factory to a training school in Florida. My total flying time for the month of March was 39 hours, 45 minutes.

In April 1944, I was stuck on the Dallas shuttle of P-51s from the North American factory at Hensley Field between Dallas and Fort Worth. This consisted of checking the ship while it was on the ground to make sure it was put together correctly; making a 30-minute flight over Hensley; flying on to Love Field, Dallas; and then an interminable wait for ground transportation back to Hensley Field to do the whole thing over again. The wait was for all of the other shuttle pilots to come in, "sell" their airplane to operations, and get on the bus. No wonder no WASP enjoyed being put on the shuttle run at Dallas!

Taking a pursuit aircraft right off the assembly line to fly it to some destination without its ever being test flown meant you were the test pilot for the first 30 minutes or so, until you found the airplane "airworthy."

I always carried a snub-nosed screwdriver in my flight suit as a standard piece of equipment. The one-piece flight suit had two pockets below the knee, right in front, and the screw driver banged against my sore shin (scars from college field hockey) all of the nine months I flew fighters. I note women pilots flying in later years have nearly the same one-piece suit ("shucks," no drop-seat), but the pockets are

Houston December weather and Kay Gott in fleece-lined suit for open cockpit flying. No gloves were issued.

sensibly on the sides where they belong, below the knee. Some other pilots used a coin, rather than a screwdriver, to open all the inspection plates and plunk your finger across all of the cables — just to make sure they were taut, and **went** somewhere.

We learned never to trust gauges, and never trust the flight-line attendant. Planes were always coming and going, with no one to keep track. It was **your** life. Visual inspection involved a walk completely around the craft before you got in the cockpit to check the controls to make sure that all the cables were hooked up correctly and the coolant wasn't on the ground. Then you climbed up on left, then the right wing, opened the gas caps and stuck your finger down to feel if there was "wet" gas there. Once you climbed into the cockpit, there were many more checks.

Remember, this was a time when people were building airplanes who had never done it before; qualified help was scarce; and inspectors were just learning their job.

Being on a shuttle run, pilots do not log very much flying time. That April 1944, I only had two other trips with P-51s to Newark, New Jersey, then a P-40 to Louisiana training school. All total, I only logged 26 hours, 30 minutes flying for April.

I had been taking P-51C models from the Dallas factory to the modification plants which were at Kansas City and back in Long Beach to the North American factory. The modification was the "bubble" canopy, which enabled pilots in combat to have unrestricted vision. The Mustang P-51D with the bubble was proved a valuable war weapon.

✯✯✯

All of May was spent at Orlando learning to be an "Officer" in the United States Army Air Corps! This never came about, so it was a waste as far as flying went! All women pursuit pilots felt the same about that "lost" month when it came their turn to go to this Army Air Force School of Applied Tactics.

Florene Miller, WASP Squadron leader at Love Field keeps everyone marching past Basic Trainers.

North American P-51.

June 6, I took a P-51C from Dallas to Kansas City. I logged 28 hours, 20 minutes flying in June. It is interesting to see that one of the trips was in a Basic Trainer, the BT-13. Our job was to move airplanes; and if a pursuit pilot was qualified to move BT-13s and they needed to be moved, that's your day's assignment!

In July, things picked up for this ferry pilot. One incident in the 48 hours logged time for July was flying a P-51C from Dallas to Long Beach for the modification to "bubble canopy." The tail wheel went flat and the airplane started pulling to the right upon settling on the runway at Long Beach. The plane and I waltzed off the runway through the grass at a very high speed, away, fortunately, from the flight line where we would go smashing through the neat long rows of airplanes parked there. When I could catch my breath and finally get the airplane stopped, here we were, sitting in the middle of the field. I called the tower and asked would they please send a jeep out for me, and a tow for the plane?

✯✯✯

"Flying is the second greatest thrill known to man; Landing is the first."
　　　　　　　　Author unknown.

Shooting landings meant practice time: either "touch and go" without stopping the aircraft, or coming to a complete stop, taxiing back to the start of the line, and heading down the runway to have another "go around." Five landings and five take-offs was standard procedure for refresher course, or renewal of skill in the delicate art of landing.

✯✯✯

I always carried a slide rule on a piece of string around my neck to compute air speed, wind, fuel and mileage. I figured these cross-country trips by arithmetic first, then checked my final figures on this circular American Airlines E6B navigation tool. I never trusted my arithmetic; and if the two figures didn't jibe, I would do the whole thing of figuring the trip legs over again, until both computations came out the same. This was the evening's homework. With the exception of weather report, especially noting winds aloft for the most favorable flying altitude, I was ready to be cleared from the field by 8 a.m. I usually was "right on" with my ETA. Homework did pay off!

✯✯✯

One lucky day I was put on orders for a ferrying trip to deliver an airplane through my home town. I picked up the fighter in Niagara Falls, New York, to take to Portland, Oregon. There were no wing tanks and no belly tank. This meant limited fuel, which also meant lots of stops. Among the rules there was a safety one that said you had to land with half-an-hour fuel reserve. So I began the long day with short hops across the United States — fuel hop to fuel hop. These pursuits only gulped high octane. And I was in a hurry to make my Idaho stop that evening. I asked if Rock Springs, Wyoming, had fighter octane and was assured they did. Eager to spend the night with my folks in Idaho, I took off from my fuel stop at Cheyenne for Rock Springs.

Bell P-59A early jet at Niagra Falls, NY factory. Kay Gott and 2 unidentified men on ground. Women on wing are K. Bernheim, E. Hiester, R. Anderson and Anna Flynn.

I landed at Rock Springs to discover they had low octane gas only. That left me two choices: keep on and risk it, or return to the last stop, tank up, and skip Rock Springs. I didn't even consider using low octane, since these Bell craft did not take gently to tampering, and had a reputation for not being kind to that sort of treatment.

Furious at having been given wrong information at the stop before Rock Springs, I was very reluctant to turn back. It was frustrating to wait and wait at each stop for refueling, as I had to do all day long. Besides, I could make the Idaho stop that night if I kept going. I jumped back in the plane, studied the map and decided I could go on.

I took off from Rock Springs, figuring that if I took a short cut off the airways, that is, off the designated flight path, I could make the next refueling stop.

Down the canyons of Wyoming, into Utah, eking out the petrol, turning on the downwind to the final approach at Hill Field in Ogden, Utah, the engine QUIT!

Since the gasoline is in flat tanks on the wings, when the airplane tipped to turn, there was no fuel reaching the engines. An airplane in front of me was cleared for landing. I called the tower for an emergency landing and the pilot ahead of me was told to pull up and go around the pattern again.

As I leveled off on the final approach, the propeller was still windmilling, the engine coughed and caught, so I did not have to make a dead-stick landing. I even had enough fuel to taxi to the refueling ramp where the gas trucks were. Thank goodness, I did not have to be towed to the ramp. There I sat on the wing, quite shaken, and waited for a jeep.

When I went in to Ogden operations, I learned that I had cut a high-ranking officer out of the traffic pattern.

Feeling very guilty, I expected a reprimand. None came.

After waiting my turn for gas, I continued on to Boise, Idaho, and landed at Gowen Field, really pushing to make Idaho, for in fighters we were required to land a half-hour before sunset. I had enough time to fly over my hometown and look down at my house. From Gowen Field, I called home, and my folks found enough rationed gasoline to come get me.

Next morning, I reported early to Gowen Field to continue on. Another rule is that you must be off the ground by 8 a.m. I got my clearance, checked the weather to Pendleton and Portland and all points between, and took myself and my flight plan clearance out to the airplane.

Good heavens!

Fluid all over the ramp under the plane!

The engine coolant had leaked during the night in the nippy desert air and was all **under** the airplane instead of in it, where it belonged. The hot engine had heated the rubber hose connections and during the cool night, the hose connections had contracted. Coolant, like Prestone antifreeze, is very viscous, and all of it just oozed out all over the ramp. Since Gowen Field had no pursuit airplanes, they had no coolant to replace what I'd lost.

Gowen Field was a B-17 base; and a squadron of WASP was stationed there to tow target sleeves over the Oregon desert to train the B-17 gunners how to shoot at a moving target. These girls were flying B-26s to keep ahead of the B-17s.

So I got to stay one more night in Idaho. They sent a B-17 down to Hill Field at Ogden, Utah, for the coolant. I called my folks to come get me again.

One of the men pilots who had left Niagara Falls with a P-63 the same time I did caught up with me at Gowen Field as they were replacing the coolant the next morning. He stopped for gas and told me he had pushed it to stay overnight in his home town, Ogden, Utah! He had planned this maneuver carefully just as I had planned to stay overnight in my town in Idaho.

As he left, he did a spectacular chandelle maneuver off the end of the Gowen Field runway, undoubtedly to impress those B-17 pilots, and then waited for me upstairs to take the lead to the next refueling stop at Pendleton. We flew down the gorgeous Columbia gorge, right over the river; but at the west end of the gorge where the gorge walls began to open up, I looked back to see him drift to the north. I called him to say that Portland was west ahead, but he said he'd never seen Mt. Ranier. After I delivered my plane at Swan Island Base in Portland and signed the papers over, he came in a half hour later.

My logbook shows I delivered this P-63 in Portland, Oregon, August 5, 1944. The next entry, August 6, I flew a P-51D out of Long Beach to Coolidge, Arizona, delivered it on August 9 to the European war port of embarkation at Newark, New Jersey. I then picked up another P-63 at the Bell factory in Niagara Falls, New York to fly to Fort Myers, Florida. I did not return to my base at Dallas Love Field until August 11.

When I picked up my mail at WASP barracks at Love Field, Dallas, I found a note to report to Headquarters. Would I get punished for cutting the high ranking officer out of the landing pattern in Ogden? Would I get the axe for running so close to "no" fuel? Would I have been reported by someone who saw me fly off the airways flight path? I just dreaded what was to come.

I guessed that I had irked the officer in Ogden, and that would probably be it, for I glimpsed him scowling at me in Ogden as I left the operations there with my clearance for Boise's Gowen Field.

"Yes, sir, Kay Gott reporting."

"You have a violation."

(No response from me. I just waited for the wall to cave in. Which error would it be?)

"You failed to fill out a clearance for one leg of your trip. Rock Springs, Wyoming to Ogden, Utah. You are suspended for three days."

It took a few seconds to digest this information. Horrors, I had not even thought of that! Good grief, was that ALL? Then I became indignant. "How many days leave do I have (had I accrued)?" (We were on civil service. We were civilians flying for

Ferrying out of the Bell Factory, Niagara Falls, NY. Men are unidentified. Women are Marion Carlstrom, Isabel Madison, Gretchen Gorman and Ruth Lindley in front of a P-63. (Ruth Lindley collection)

the Army Air Corps, and living on the base and taking orders from the AAF.) I told this man that I would add the three days suspension to my days earned, and just take a vacation home.

"Oh, no. This is punishment, and not to be used for vacation!"

So I spent the three days doing Christmas shopping in August and enjoying the city of Dallas. I was not confined to base, barracks or, to my relief, not thrown in the brig. Three days passed, but not in the air. On the fourth day, August 14, I was on orders to take a P-51C for the China account from Dallas to Newark, New Jersey.

★★★

When weathered in, I walked a lot. I walked and walked and walked, exploring the area.

★★★

An instrument rating was required in pursuits also. Then when we got the Instrument ticket we were never permitted to fly instruments with pursuits at Dallas, ever.

Also at Dallas, if you came in one minute **before** midnight, you had to be on orders the next morning. Operations used to watch the time that the airlines got in. If you wrote the wrong time down they were apt to call you up at 8 o'clock in the morning and say, "Listen, we just happened to check on American Airlines and we know that you came in before midnight, get up, get out there on the line."

★★★

K: I stayed to the very end, and it was such a traumatic thing to get dismissed, disbanded, when we were so useful, so needed at Dallas. It was such an unjust thing, to me from my point of view.

BJ: It seemed so, to me. Of course, they had so many war-wearies back then.

K: Yes, but remember what Colonel Higgins, C.O. at Dallas said: "I have these 33 girls at Dallas, and they are all qualified, and they all do a good job, and when I send them with a plane they all go there and they return. They do it. They don't end up in Salt Lake City when they are supposed to be going to Newark (referring to an actual incident by a male pilot)."

K: We had three sets of wings. ATC, the diamond WASP, and our Class Wings engraved with "43-W-2." Colonel Higgins had a party when we disbanded at Dallas. I got a present that every WASP got — a compact with my name on it... Lois K. Gott. We paid dues into a WASP Squadron fund, and the compacts were purchased from our dues. Colonel Higgins thought a lot of us.

K: Did you know that "Pete" Madison (Isabel) and I took the last two P-40s that went to General Chenault?

BJ: Did you?

K: The only trip I ever got to Oakland.

M: Really?

K: And the factory people wrote their names all over the planes; it was quite a ceremony to send us off, the last two P-40s, the last order. We always wondered if they ever got to him, because Chenault closed their China operations shortly after that. (We delivered those planes on September 9, 1944.) When I came into the Bay area, I flew over Alcatraz. I knew you were not supposed to fly over a Federal prison, but I circled around it.

★★★

I served two years plus a few days — eleven days.

P.R. photo with Cappy Vail, BJ Bachman, Florene Miller and Betty Whitlow by the nose of a P-40.
(Cappy Vail collection)

From WASP yearbook 43-6 and 7.

Marion Carlstrom with her check pilot and husband to be Capt. Carl Trick.

Isabel "Pete" Madison in P-51D cockpit.

Betty Whitlow. "Bee Whit" delivered many AT-6's and went on to pursuits.

Ross Kary

Ross Kary was a graduate of the March 9, 1944 Palm Springs Pursuit School with Kay Gott, Byrd Granger, Marge Ketcham and Ruth Dailey.

Ross logged many hours in the P-51, since it was made at the North American Aviation (NAA) factory some 30 minutes flight from Dallas. Often the WASP were put on the shuttle run from NAA to Dallas. It was not a good task, since there was no transportation to quickly go pick up a plane, and no easy way to return to NAA for another. Two flights a day were usual: one in the morning run, and one in the afternoon.

Ross had a dramatic time during one P-51D flight when the bubble canopy jettisoned on take-off, quite unexpectedly and without her help!

She brought the plane around the pattern and landed without further incident.

Ross Kary

Author's note: That's where Ross Kary almost lost her head. She just got up in the air, and the latch came off, and the only thing that saved her from being beheaded was the canopy hit the gunsight and flipped up. She would have lost her head at 200 miles an hour. "Rosie the Riveter" had not had the work inspected. We were test pilots.

(A cartoon sketch by A. J. May from her book *RON*.)

Vi Thurn

Vi Thurn shares her feelings in this interview:

You can't really talk about what it really is like to fly. Your audience will have no idea in the world what you are trying to talk about. So that's why, in the disappointment of being disbanded and sent home before the war was over, I didn't talk about it for years. Leaving flying was one of the toughest places in my life. I mean, it was as though someone had taken away my food — flying, the one thing I could do well — wanted to do, **couldn't** do, and there was no avenue, no reconsideration, no job for us, and no way I could survive. So I just didn't talk about it. Now I feel young people should really know about us. I didn't realize at the time that we were a part of history, and that we **did** open up the field to women now.

These women flying for the military now really appreciate us. Yes, they do!

Violet C. Thurn on the wing of a P-51C. She stretched to reach 5'2" and always carried extra cushions so she could reach the rudder pedals.
(WASP historian collection)

Gayle Bevis

I went to Brownsville to Pursuit school, and graduated June 30, 1944. We checked out in P-39s, P-40s, P-47s and P-51s. We did do transition to heavier aircraft in Dallas. I never got to the P-38. We did do a lot of P-51s, since I was stationed in Dallas. I never flew P-39s again, only did one P-40 and a couple of P-47s.

I was checked out in the P-63 Bell at the Niagara Falls factory, and they had auxiliary wing tanks hanging on them for gas. One already had a red star; it was for the Russians and I was to take it to Great Falls, Montana. I had done the empty-tank landings the day before, and I was checking out the fuel-tank landings. I did something very stupid. I used the toggle switch to put my gear down and, thinking it was all the

Gayle Bevis

Gayle Bevis stands on the wing of a P51D. Note the headlamp under the wing. (Gayle Bevis collection)

way down, put the switch in neutral. But the gear was not all the way down. The main gear was within two or three degrees of being all the way down; but they held. The nose wheel, unless you are down and locked, collapses on you; and it did!

I landed, and was skating down the runway on my nose when I looked up and thought my auxiliary tanks were dragging on the concrete. I thought I would be sharp and make a fast decision and look good in Dallas if I got out of the airplane fast before it blew up, so I opened the door.

It turned out later on that I had a four-inch clearance between the nose of the plane and the runway.

Since the P-63 has the engine behind, the cockpit sits so far forward on the wing that probably when I went out I hit the ground running off the back of the wing, which I intended, but I ran over myself with my own airplane.

I ended up in the hospital with a fractured ankle and skull apparently from being hit by the tail assembly. They put me in a walking cast. The skull fracture was an incomplete linear fracture of the left occipital bone. That means I just cracked it and didn't go all the way through.

Niagara Falls sent word to Dallas, and they sent their own C-47 for me at Utica. I went to Mitchell Field and was put on a real hospital plane with overseas casualties.

So I kept a very low profile as I tried not to get seen too much, and we made stops all the way back to Dallas.

I was there for the finish (disbanding) on December 20, 1944, then was sent to General Hospital in McKinney, Texas.

Gayle's P-63s nosewheel collapses. (A cartoon sketch by A. J. May from her book *RON*.)

(Author's note: In the Dallas final-day photo, Gayle is in a leg cast.)

★★★

A few weeks later, I went home and recovered. I wasn't in too bad shape, really, after a week. That was my last flight. I missed out on the last six weeks of flying since this happened on November 12. After I recovered, I was the last one in from Bryan, Texas, and the two pilots let me fly the airplane. I didn't do very well with my walking cast.

Helen Turner

I delivered a P-51 to Newark on November 11, 1944 and, to my surprise, found the administration building full of pilots, both men and women, milling about uncertainly. I signed over my plane and joined the crowd.

Toward dark, we were called together and received some explanation for this. We were to be flown to Niagara Falls, where we would wait for further orders at the Bell Aircraft plant.

Next morning at Niagara, we were introduced to Bell's new pursuit plane, the P-63. An instructor briefed us on the finer points of the plane and sent us up for check rides around the area.

This brief flight was the beginning of my most eventful trip in the ferry command — and the saddest.

I had flown for half an hour or so and was preparing to make my final approach when the tower warned me off. A plane landing ahead of me was in trouble. The nose gear had collapsed and now the plane was tumbling down the runway.

When I landed and entered the administration building, my name was being called. I was told Gayle Bevis had been the pilot in the tumbling plane.

When the nosegear collapsed and her plane started to tumble, Gayle had opened the door, as she had been instructed to do in such circumstances, and jumped. But she jumped too soon. Her

Helen Turner at the wheel well of a P-51D. (Helen Turner collection)

plane had still been moving at 60 miles per hour, so Gayle was badly injured.

I was wanted when I arrived because no one knew where Gayle was based or how to contact her family. Also, no one was sure of where the ambulance had delivered her.

One of the pilots and I agreed to look for her. We caught a taxi and tried several hospitals. Finally, at about 9:00 p.m., we found her still unconscious in a hospital bed.

Over and over I asked her to tell me her mother's name and where she lived. I had little faith that an unconscious person could understand me; but eventually she replied, and I called her mother in Kansas City. Gayle remained unconscious for three days after that, I'm told, but she did recover.

The next day I wasn't permitted to wait for word about her. I was sent on my way.

Gayle returned to Love Field, and I've seen her since those days.

Helen Turner is at the controls of a P-51D in flight. (Helen Turner collection)

Ruth Trees wears suntans and has a parachute slung over her shoulder as she stands by a trainer plane. (Ruth Trees collection)

Jane Emerson is geared up and at the controls of a P-38, ready to go. (Jane Emerson collection)

Women in Pursuit

Helen Schaefer stands on the wing of a P-47. Note her parachute "bustle". (Army Airforce PR photo) (Helen Schaefer collection)

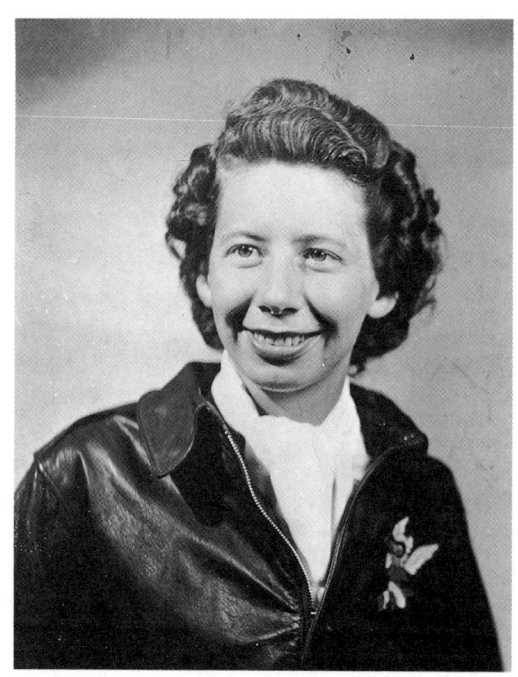

Helen Schaefer

Monthly totals from Helen Schaefer's Flight Log (Logbook). She speaks proudly that she has one hundred hours in the P-51 Mustang - her favorite pursuit plane.

```
                     HELEN SCHAEFER
                                            Primary time: ....50:02
Entered WASP training February, 1943        Basic ........... 70:08
in WASP class 43-W-4                        Advanced ........ 50:07
Graduated August 7, 1943                    TOTAL ...........170:17

Assigned 5th Ferrying Group, Love Field, Dallas, Texas

     August 1943.......................      0
     September..........................  12:15   plus 1:55
     October............................  35:40
     November...........................  51:00   BC-1
     December...........................  28:45   AT-6, UC-78
     January 1944.......................  13:30   UC-78, BT-13
     February...........................  22:15   AT-6, AT-7, UC-78
     March..............................  46:05   UC-78, AT-7, PT-13, BT-13,
                                                  AT-11, RA-24
     April..............................  34:15   AT-11, RA-24B, C-47, UC-78 Link
     May................................  53:45   C-49X, UC-78, Link, C-47,
                                                  AT-6, C-53
     June ..............................  00:00   Orlando, Florida. Began June 6
                                                  for 30 days, OTS.
     July...............................  38:50   AT-7, Link, AT-6
     August ............................  30:35   AT-6, BC-1, Link, P-47, P-40,
                                                  P-51, P-39, P-63

Pursuit class 44-16B, graduated from Pursuit
School, Brownsville, Texas August 31, 1944

     September..........................  34:05   P-51, P-40
     October............................  41:45   P-51, Link
     November...........................  31:45   P-51, Link, P-63, AT-6
     December...........................  26:45   P-51, F-6K
                                                  Last trip December 15, P-51D
                         ................675:05   time to December 20, 1944
                                                  WASP deactivation date
```

Betty "BJ" Bachman

BJ: Yes, I was in pursuits. I was just told I was going to pursuit school, and one afternoon, without a clean shirt or anything, they put us on SNAFU for Palm Springs and school. I was in Palm Springs when Dorothy Scott was killed. They thought I was the one killed, because I was supposed to have been up at the time. People looked at me strange.

✯✯✯

BJ: I have my class wings. They say "319" on the top.

A lot of the gals had uniforms; I never did get one. I always wore pinks and greens, and suntans in the summer.

✯✯✯

There were so many guys compared to us.

Betty Jane Bachman

The 5th Ferry Group WASPstand to colors in Dallas on December 20, 1944. Back row, Kay Gott, Helen Schaefer, Isabel Madison, Evelyn Tomlinson, Helen Abell, Jan Tate, Virginia Luttrell, Connie Colby and Gayle Bevis (still wearing leg cast from her accident). Front row, Sidney Miller, Virginia Harris, Vi Thurn, Marion Carlstrom, Gretchen Gorman and Nelle Carmody. Col. Albert Higgins addresses Ruth Dailey.

BJ: When I reported for training to get my physical, the poor Captain had never had a girl, and I was just barely stretching it to 5'2"

Kay: With all your fantastic flying, you would not qualify to fly later as Jacqueline Cochran recommended at the conclusion of the program that women should be 64 inches.

BJ: Actually, I think they should, because if we got in a P-47 we disappeared from sight.

K: How many pillows did you carry? And Vi Thurn is smaller than you.

BJ: Depended upon what I sat. I carried a little overnight bag, and my parachute pack and another cushion.

★★★

BJ: You weren't supposed to land at Wright Field in Dayton, Ohio, during the war either, but I did. Have you ever been met at the plane with submachine guns, and kept at the field?

★★★

BJ: You know, they sent me over to Grand Prairie, the North American P-51 factory at Hensley Field near Dallas, right after I completed pursuit school at Palm Springs. Right after I graduated I had my first P-51C to Newark, and I came back and I was sick in bed. The Army wouldn't give me any help, I had to go to a civilian doctor, who promptly put me in the sack. I got over this pneumonia, and the first thing the Base did was put me on orders to go on this shuttle. So without any checkout, I climbed into the P-51 over there at Hensley and taxied it out, and we were on the short runway over the lake, very short runway; and they were in a hurry, so I gave it an awful lot, and not having the feel back at all — you've got to wait for me — that was a mistake, I'm sure, that will be remembered to this day. They had never seen an airplane act this way. It veered off over here, it got off on another runway, tractors and earthmoving equipment all over the place; I leap-frogged those.

Marie: Did you ever get off?

BJ: I got off eventually. I think it did it itself; I was just cross-eyed. So I flew around an hour, just getting the feel of the airplane, and saying to myself, "I can't put this on the ground if I can't get it off."

M: If you are away from it for awhile, you need to get practice.

BJ: Right. And that P-51 had so much power. I hadn't allowed for the torque and I hadn't adjusted the trim.

★★★

BJ: Maybe I should have stayed in until December 20, 1944, and put off this wedding. I was sick though. I'd had pneumonia. You know, I got out in Palm Springs. I flew for a long time, feeling very ill. I kept getting thinner and weaker, and then I began to have so many little things happen to those P-51s — that you get a feeling about this sort of thing. It's time to quit.

★★★

K: Where did you go to Instruments?

BJ: Love Field, Dallas. I got all my Link in Houston. I loved instruments. Do you know we hadn't had but two hours in Link in Houston when Ginny Alleman and I were sent out on cross-country in AT-6s. We went directly west. We made our first check point and then I lost Ginny. Heading for the second check-point, the weather closed in, and I didn't know where Ginny was, and I had only two hours on the Link, and I had to work an Instrument problem, because it was completely socked in back to our training base at Houston. I know at one time I was over the Gulf, as I could see an awful lot of water through one break in the sky. So I worked the problem and found the Cone of Silence and found Houston, and broke out of the clouds, and it was dark! And just about that time Ginny broke out. All of the instructors were out on the field and we came in and landed where Ginny landed, she was so glad to get on the ground she stepped on the brakes too hard and set it up on its nose!

M: Oh!

BJ: Just as I was taxiing up to the flight line to turn, I ran out of gasoline. That was wild! Ginny had had to do the same problem, that is, find out where Houston was.

Attending a buffet supper at the WASP Farewell Party, given by Colonel Higgins at Love Field on December 19, 1944 were Dorothy Hopkins, Rosa Lea Fullwood, Virginia Luttrell, Marion Carlstrom, Barbara Willis, Virginia Harris, Ruth Lindley, Nelle Carmody, Helen Abell, Isabel Madison, Evelyn Tomlinson, Helen Schaefer, Hazel Pierce, Norma Jane Emerson, Delphine Bohn, Margery Taylor, Esther Nelson, Jean Taylor, Lucy Walker, Ann Johnson, Magda Tacke, Ruth Dailey and Gretchen Gorman.

M: Coming out of it, too. Think of all the close calls you had. We all had.

BJ: We all had sessions with the Flying Safety Officer: "Why didn't you have this taken care of?" Ginny and I were in trouble for a long time.

Dorothy Hopkins, Rosa Lea Fullwood, Virginia Luttrell, Marion Carlstrom, Barbara Willis, Virginia Mullins, Ruth Lindley, Helen Abell, Isabel Madison, Evelyn Tomlinson, Helen Schaefer, Hazel Pierce, Jane Emerson, Delphine Bohn, Margery Taylor, Esther Nelson, Jean Taylor, Lucy Walker, Ann Johnson, Magda Tacke, Ruth Dailey, Gretchen Gorman and Mary Edith Engle receive gifts from Base Commander Colonel Higgins.

Helen T. Abell

Marion V. Carlstrom

Nelle Carmody

Cornelia Colby

Helen Irene Fremd

Rosa Lea Fullwood

Gretchen Johnson Gorman

Dot Hopkins

Ruth E. Lindley

Virginia D. Luttrell

Isabel Madison

Sidney Miller

  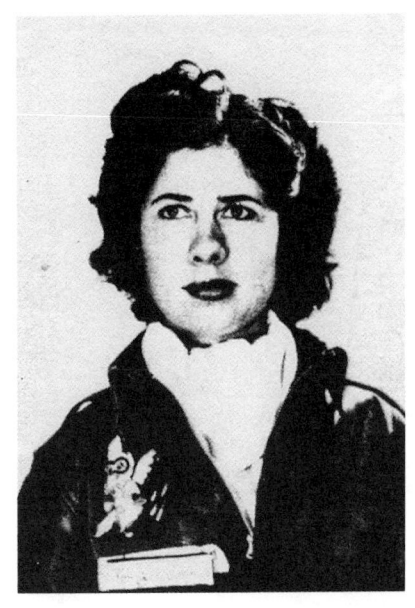

Janice R. Tate   Evelyn Tomlinson   Gertrude V. Tompkins

Betty Whitlow   Barbara Willis

Mixed emotions show on the faces of the 5th Ferrying Group WASP on the day of disbandment, December 20, 1944. Standing Delphine Bohn, Sidney Miller, Frances Dias, Ruth Dailey, Esther Nelson, Janice Tate, Evelyn Tomlinson, Helen Schaefer, Virginia Luttrell, Isabel Madison, Magda Tacke, Jean Taylor, Lucy Walker, Nelle Carmody and Margery Taylor. Kneeling, Norma Jane Emerson, Ruth Lindley, Gayle Ewing, Gretchen Gorman, Helen Turner, Irene Fremd, Connie Colby, Barbara Willis and Marion Carlstrom. Seated, Ann Johnson, Virginia Mullins, Dorothy Hopkins, Hazel Pierce, Helen Abell, Rosa Lea Fullwood, Kay Gott, Violet Thurn and Mary Edith Engle.

6th
Ferrying Group

6th Ferrrying Group
Air Transport Command, Long Beach, California

Name	WASP Class	Pursuit School	Graduation Date
Beaty, Lauretta	43-W-4	44-9B	5/15/44
Coleman, D. Lewise "Lew"	43-W-2	44-2	1/44
Cummings, Iris	43-W-2	44-1	12/43
Cusack, Lana Blanche	43-W-6	44-12B	6/30/44
Erickson, Barbara Jane "BJ" (Squadron Leader)	WAFS	—	
Featherhoff, Nancy L.	43-W-6	44-12A	6/30/44
Fillmore, Carole	43-W-2	44-2	1/44
Fisher, Enid C.	43-W-6	44-12A	6/30/44
Harris, Thelma Nadine	43-W-7	44-15B	8/15/44
Hill, Virginia "Ginny"	43-W-4	44-10B	6/1/44
Kocher, Dorothy "Dottie"	43-W-4	44-10A	6/1/44
Landis, Jean	43-W-4	44-13A	7/15/44
Leaming, Kittie	43-W-4	44-11A	6/16/44
Loft, Katherine "Katy"	43-W-4	44-10A	6/1/44
Miller, Florene	WAFS	44-2	1/44
Nichols, Dorothy "Dottie"	43-W-2	44-9A	5/15/44
Ramsey, Nadine	43-W-5	44-8B	5/1/44
Richards, Helen	WAFS	44-2	1/44
Rutherford, Madge A.	43-W-4	44-11A	6/16/44
Scharr, Adela R.	WAFS	—	
Sharp, Evelyn	WAFS	—	
Tackaberry, Betty C.	43-W-1	44-8A	5/1/44
Thompson, Ruth R.	43-W-2	44-1	12/43
Trammell, Evelyn L.	43-W-6	44-13B	7/15/44
Truax, Deborah "Debbie or Deb"	43-W-6	44-13B	7/15/44
Webb, Dorothy E. "Dottie"	43-W-6	44-12A	6/30/44
Wilkes, Rena D'Arcy "Billie"	43-W-4	44-18B	5/15/44

Barbara Erickson

Barbara Jane Erickson was in the first primary non-college CPT course in Seattle, and went on to earn her secondary, cross country, instructor and commercial license in the CPT programs. At completion of these, she started instructing in Seattle, entered the WAFS in September 1942 and served with the WAFS and WASP until they were disbanded on December 20, 1944.

✯✯✯

On March 11, 1944, a little-heralded event took place at Sweetwater, Texas. At the graduation class ceremonies of Class 44-W-2, General "Hap" Arnold, Commanding General of the Army Air Forces, awarded WASP Barbara J. Erickson the Air Medal in recognition of outstanding service.

An accomplished pursuit pilot, Barbara was WASP Squadron leader at the 6th Ferrying Group, Long Beach. Much to her surprise, someone in Air Transport Command inquired and was concerned that she was overworked, since she had just delivered five aircraft in four days, coast to coast.

..."certain male officers persisted in believing that women were constitutionally too weak to engage in long ferrying flights.

CPT instructor Barbara Erickson is being congratulated by Air Force General "Hap" Arnold after graduating six pilots. Robert Hinckley is on her left. 1941

Barbara Erickson in cockpit ready to go, as Evelyn Sharp, standing on the wing, checks out sky condition. (Barbara Erickson collection)

The whole matter of fatigue was discussed between ATC and AAF HQ as result of a routine mission performed by Barbara Jane Erickson — one of the original WAFS, who in slightly more than five days of actual flying, made four 2000-mile deliveries. This mission was reported to ATC in the routine weekly report submitted early in August 1943. As result, much to the astonishment of the Ferrying Division and Miss Erickson herself, AAF Headquarters directed ATC to initiate a recommendation for award of the Air Medal to Miss Erickson.

"The latter tried to refuse the honor, since she felt she had simply performed an ordinary mission and should not be singled out from among the other women pilots for special honor. But she was directed to comply and, in a formal ceremony at Sweetwater (which school she had never attended, being an original WAFS), the medal was duly awarded in the presence of Miss Cochran, Mrs. Love, and highranking generals of the Army Air Forces, including General Arnold who presented the medal.

"After the curious affair in which a woman pilot was given a medal for doing the job for which she had originally volunteered, AAF HQ became concerned over the possible mental and physical strain imposed upon Miss Erickson by assigning her so many missions in so short a time. The question was asked of ATC, 'Why was this mental and physical strain imposed upon this pilot?'

"The question was answered by ATC HQ by giving a simple analysis of the flights: the five flying days were not necessarily consecutive. She left Long Beach, California, July 29, 1943 and delivered a P-51B to Evansville, Indiana. This was a daylight trip. After a night's rest, she left Evansville July 30 and delivered a P-47 to San Pedro, California, July 31st, resting overnight on the way. It is a regular practice for Ferrying Division pilots, male and female, to be assigned ferrying missions by special control officers stationed at factories and Modification Centers, when these pilots have made deliveries and are about to return to their home stations. This avoids wasting time by riding as passengers back to their home stations. Male and female pilots are often gone on a whole series of

Barbara Erickson WAFS, Squadron leader of 6th Ferrying Division, based on Long Beach, California. (Barbara Erickson collection.)

ferrying missions for so long as a month or six weeks before returning to their stations.

"In the case of Miss Erickson, if she had not flown a P-47 back to San Pedro, she would have had to sit up all night as a passenger on an airline. By ferrying a P-47 she was able to get a night's sleep in a hotel.

"Miss Erickson's third trip began on August 2nd from Long Beach when she ferried a C-47 to Fort Wayne, Indiana. She was then on duty from August 3rd to August 6th at Ferrying Division Headquarters in Cincinnati. On her trip home, the Control Officer at Evansville again assigned her a plane, a P-47, to deliver to San Pedro, a short distance from her home station. In view of these facts, ATC HQ did not believe that any extraordinary mental or physical strain had been placed upon Miss Erickson.

Cy Younglove, Flight and Field Operations Superintendent, talks with Barbara Erickson. The P-38s in the background are at the Nashville, Tennessee Lockheed Division for modification in 1944. (Barbara Erickson collection)

General Hap Arnold presents Barbara Erickson with the Air Medal on March 11, 1944 in recognition of outstanding service. (WASP Historian collection)

"ATC HQ had also been asked if other pilots, male or female had been available. ATC explained that they were available but that pilots are kept on a rotation list when they report back to their station, in accordance with their pilot classification. At this time Miss Erickson was a Class IV pilot and she was next in line to make a delivery. ATC then pointed out that, 'In the strict sense of the word there was no emergency condition. The activities engaged in by WASP Erickson were normal considering the number of domestic deliveries made during the period she was flying."

In summary, Barbara flew 8000 miles in the course of ten days. From Long Beach, California, she flew a P-51 to Evansville, Indiana on July 29, 1943; the next day, a P-47 to San Pedro, California, delivering it on July 31st. On August 2nd she left Long Beach flying a C-47 to Fort Wayne, Indiana. She proceeded to HQ Ferrying Division in Cincinnati, Ohio, for temporary duty. On August 7th, she reported to Evansville, Indiana, picked up another P-47 and delivered it to San Pedro, California. Thus, out of ten days she was able to spend five in ferrying. Estimated logged time was at least forty hours, as much as one of the Wilmington 2nd Ferrying Group WASP could hope for in a full month.

She accepted the award in the name of the WASP, and for all the WASP. To her, this was not that unusual an event. Other WASP also served well.

Katherine S. Loft

Katy Loft

Katy Loft, WASP class 43-W-4, based at the 6th Ferrying Group in Long Beach wrote in a letter dated August 30, 1988:

As for choice of pursuit school or not, the girls in the Ferrying Division were expected to ferry pursuits. It is my understanding that those who washed out of pursuit school were transferred out of the Ferrying Division. At Long Beach, we had P-51s, A-20s, P-38s and P-61s to ferry. Most of us were checked out in C-47s before pursuit school. I delivered two C-47s, then went to St. Joseph, Missouri, for Instrument school, and also had a B-25 check-out.

I know some of the girls didn't like flying the pursuits and resigned, but I loved it.

Florene Miller

BJ: We have to rehash Florene flying through all the light wires when she blacked out Love Field in a P-47. That's why she was disciplined. And she did not get to go out to pursuit training with the rest of us.

Unbeknown to the Group Commander, one of the boys who was a pursuit check-out pilot at the Love Field base checked her out in the P-47 at Dallas. We had not been authorized to fly in pursuit. Anyway, he checked her out. So she was up. She had made several landings and flown around. Just toward dusk she came in low; it started sinking on her.

Kay: Pilots nicknamed the P-47 Thunderbolt "Bucket of Bolts," it was our heaviest fighter plane.

BJ: She hit a telephone pole — she was trying to land close to the end because we were not too sure about the range of

Florene Miller wears a winter flight jacket, scarf, flying helmet, and goggles. Note ID tag and ATC symbol, October 1942. (Florene Miller collection)

these pursuits — on landing they didn't stop as quickly as the others. She came in low, and the telephone pole which carried all this high voltage equipment came in right back of the coolant radiator on this P-47 — Jug — lit there, ripped the entire belly off the ship, clear back, and it started to flip. Florene is a good pilot. She righted that thing and gave it the gun; how she did it, I don't know — she got up and around. She lost her receiver but she could still transmit. The receiver on the radio was out, the flaps were out, the gear was out. She was full of gasoline. She did manage to pick up Meacham Airport at Fort Worth. Anyway, they could raise her and Grand Prairie could, but the Dallas airport couldn't. So she flew around for a few hours.

K: In the dark?

BJ: Fran Dias and I went over to the control tower. Now, I knew the guy from the tower since I had worked for Braniff, so

Florene Miller, radio earphones in hand and parachute pack in place, prepares to enter the cockpit of an AT-6 at Love Field, Texas in 1943. (Florene Miller collection)

Florene Miller checks out of the cockpit of a P-51 at Long Beach, California in 1944. (Florene Miller collection) (Authors note: This photo appeared in color on the cover of the spring 1944 North American Magazine)

we got up in the tower and sweated it out. Everything was dark! Of course, the tower was out — the landing lights, everything — all the lights on the field were out. We lined up jeeps.

K: Could you see her?

BJ: No, but we could hear her occasionally. She flew around — it must have been two and a half hours. Her radio was damaged and we could hear by relays, either Grand Prairie or this other field.

K: What was she doing? Was she circling?

BJ: Yes, we got messages. She was circling; and she flew out and circled over White Rock Lake, because she was afraid she was going to have to leave the plane. She packed everything in the little pockets in her pants, and took all the ship's logs and flew out over the lake to check it out. She found out she could fly the thing; she found out she had some maneuverability. She cranked the gear down and it seemed to hold. So she thought she would try to land it, but they were trying to talk her into jumping. She didn't want to — she was more afraid of jumping than she was of landing; it was dark and she didn't want to land in the water out there. So eventually she came in. They lined up jeeps on either side of the runway, and turned on the jeep lights, and that lighted up the runway for her. Meanwhile, all the commercial airline flights into Dallas were circling, holding in pattern up there, and she came in and landed that thing! It was a sloppy landing, but she got it on the ground! She taxied in, and was met by the Commanding Officer. The fellow that checked her out was a Major — he ended up a 2nd Lieutenant on this deal. As a result of this episode, Florene did not get to go to pursuit school in the first class. We were about to leave in a few days.

K: And she was relieved of WASP Squadron Command, and Delphine Bohn took over at Dallas. Florene was transferred to Long Beach, 6th Ferrying Group.

Marie: What was the punishment for Florene? What did they do to her?

K: Well, they shipped her out.

BJ: She was not allowed to go to that class in pursuit.

M: Did she go to pursuit?

BJ: She did go later. She couldn't go in the first class.

M: I see.

BJ: She came in the second class.

★★★

Iris Cummings

Iris Cummings, Palm Springs, California, Pursuit Class of December 1943, relates:

I was in one of the first classes of Pursuits at Palm Springs in December 1943. I qualified and flew B-25s and A-20s. I was flying P-38s and P-61s, the Night Fighter (Northrup's Black Widow, the U.S.A.'s faster fighter in WWI), and delivered them from factories near Long Beach where I was assigned to the 6th Ferrying Group. Upon deactivation of the WASP, I received my FAA Federal Aviation Instrument Rating at Long Beach flying with an FAA Inspector.

Lewise Coleman

Lewise Coleman, 43-W-2 WASP class, was based at the 6th Ferrying Group Long Beach.

My pursuit class was the first pursuit class with women, along with Iris Cummings. My pass (identification) for Palm Springs is dated December 1, 1943 and my diploma is dated December 30, 1943. I received my instrument training on April 3, 1944, so I could not have delivered P-38s

Iris Cummings, P-38 pilot. (Lewise Coleman collection)

before then. I took Instrument transition right there in Long Beach, California, using any twin engine that was available — B-25s, A-20s, C-47s. Took my instrument flight check final on a B-17; as I recall, there were two men pilots being checked out on the same flight. I do not remember which WASP flew what airplane, except that Carole Fillmore and I were often on P-38 deliveries at the same time. I was usually alone.

I came across my copy of "Operating Instructions for the P-38 H and J airplanes, dated 16 October 1943." One section was Emergency Gear and Flap Lowering Procedure — 3 1/2 pages long. I think I told you that I had to pump down the gear once and made an unscheduled landing at Monroe, Louisiana. But the point I want to make was that not once when I was flying the P-38 do I recall worrying about a failure occurring. From the book, though, it sounds as though

Iris Cummings and Lewise Coleman, 1943. (Lewise Coleman collection)

Lewise Coleman in full uniform.

Lewise Coleman standing on the wing of a P-38. (Lewise Coleman collection.)

it happened **often** in that ship and that we could almost **expect** it to happen some time!! Interesting!

I did fly many P-51s to Newark, then went to Niagara for P-63s or P-39s and on to Buffalo for P-40s; sometimes I picked up a P-47 in the East. But after starting on P-38s it was usually home immediately to pick up another P-38 and get on the way again. I was so involved with my great fortune that I didn't pay much attention to who was flying what.

My copy of "Pursuit Training Program" lists us as Class 43-1 (pursuit school) and gives 45 hours of lectures, films, discussions, explanations in classroom with mock-ups and on the flight line. The exams came every four or five days. The flight schedule shows 24 days of flying — a total of 47:30 hours. I don't think we did the seven hours of "formation" flying though.

Dorothy Nichols

Dorothy Nichols was a graduate of UCLA with a Master of Arts Degree. Her hometown was nearby Van Nuys, California. She was Associate Professor of History at Louisiana State University, Baton Rouge, Louisiana, where she took Civilian Pilot Training and obtained her Private Pilot's license.

After the war began, she took a leave and became Inspector for Chemical Warfare Service in San Francisco. She was active in women's athletics and kept up her interest in flying.

She entered the Women's Flying Training under Jacqueline Cochran, and graduated with the second group of women in this program on May 28, 1943.

Dottie was assigned to the 6th Ferrying Division of the Air Transport Command at Long Beach. After a year ferrying lighter aircraft and upgrading her skills, she obtained her Instrument Rating. She was sent to Brownsville, Texas for pursuit school, and graduated from there on May 15, 1944. She qualified as pilot in P-47s, P-40s, P-51s and P-39s.

Dorothy Nichols

Almost a month after graduating from Pursuit School, she was killed while en route to deliver a new Bell Aircobra P-39Q from the factory at Niagara Falls, New York, to Great Falls, Montana. At the refueling stop at Bismarck, North Dakota, on the take off leaving the field with a full belly tank of fuel, her engine quit and she perished.

★★★

The Bell factory sent a representative to Bismarck after these June disasters. He would meet the Bell airplanes as they taxied to the line, and he remained with the planes as the pilots refueled. One supposition about the presence of the Bell representative was to deter further acts of sabotage.

★★★

Juanita Bolish of the 3rd FG talks about her Flight B classmate Dottie Nichols.

All I can tell you is that she asked me to go out on a date with her the night before and I didn't go. That next day she took off and the engine quit on take-off. This was at Bismarck and she burned. I saw them through the hotel window when they had her on a stretcher. She was covered up with a blanket, but her knees were bent up. That's what I remember.

Kay: Did you ever hear any reason why her engine quit?

Juanita: Well, we were all grounded at Bismarck, because we had so much trouble. They suspected sabotage, you've heard this... at the factory, and said there had been sugar put in the gasoline. But that's all I know about it. I had to stay there two weeks because they had to go through every plane and check it out. I went out to the end of the runway to take off, and I was so scared that I got out and vomited all over the tarmac. The sergeant had to come out and start the plane. He gave me his beautiful clean handkerchief and I took it with me.

K: We all thought it was sabotage, but it was very hush-hush.

J: Well there was never anything official on it for that reason, because another... I am afraid to say this, that it was the same flight, but I think it was... that the fellow took off and turned back and there was smoke coming out... it was 44 years ago.

K: This was not the same time that Betty Shea jumped out of a plane?

J: No. I don't think so.

★★★

Martha Wagenseil, 21st FG, was also at Bismarck on June 11, 1944.

Martha: I watched Dottie Nichols die.

Kay: Tell about that.

M: I don't remember what the date of this was, and I don't remember if it was Fargo or Bismarck. I know a group of us in P-39s or P-63s — I think it was P-63s ...out of Bell Aircraft.... there must have been about eight of us, taxiing out together to take off. Dottie Nichols was first in line. I was second in line, and a guy who had been through pursuit school with us in Brownsville, a round-faced fellow, I can't think of his name, was third. We were all acquainted with each other.

Dottie ran up her engine in front of me. We were nose to tail on the ramp there to the head of the runway, and her engine was so bad and so rough, and so sputtering that those of us behind her pulled off the taxi strip before she moved to give her room to taxi back to the flight line. To our amazement and despair, she **didn't** taxi back to the flight line. She pulled onto the runway and took off. She rolled down the runway, lifted into the air, and got about ten feet in the air when her engine quit. She was carrying a big belly tank... this airplane was going up to Alaska and then to Russia. We had red stars painted on the fuselage... they were ready to go to Russia. She landed straight ahead... which was absolutely right. There was a field ahead of her, and she went up and she came down. She handled the airplane well. She landed straight ahead, but she landed on the belly tank.

And, at the point of impact, the aircraft and the belly tank separated from each other. The belly tank went skidding down the field, the P-63 itself bounced into the air and slowly settled back down on **top** of the belly tank and the whole thing blew up.

I later was told by someone on the accident investigation committee that Dottie did not die of burning in the crash; her head had snapped forward and the control stick had gone through her forehead. You felt sure that she had died before the body was consumed.

Lauretta Beaty

Lauretta Beaty, WASP 43-W-4 class, was based at 6th Ferrying Group, Long Beach, California.

Lauretta worked for Warner Bros. in Hollywood appearing in movies with stars such as Esther Williams. She also worked at a Piper Aircraft factory before joining the WASP.

"I picked up P-39s and P-63s at the Bell factory in Niagara Falls, New York, and flew them to Great Falls, Montana, where I signed them over to the Russians. You can tell it was cold. I have a jacket and sweater and ski underwear under that flight suit,

Lauretta Beaty with P-63 for Russian Account (Red Star on side) and wing tanks. Winter in Niagara Falls at the Bell factory.

and I never wore gloves unless it was very cold. Judging by the wing tank, it may be the four-bladed propeller P-63."

"The P-51 was my favorite airplane, and I delivered more of them than any other airplane."

Nadine Ramsey

Nadine graduated from Pursuit School on May 1, 1944 and was assigned to the 6th Ferrying Group in Long Beach, California.

"I bought a war surplus P-38 (total 22 hours air time) in Kingman, Arizona, after the war, and flew it home where I was met by newspaper reporters. Later Life Magazine sent out a reporter and photographer, and my picture with plane appeared on the cover of the magazine as the only woman in the world to own her own P-38 Lockheed Lightning. I flew it about 2-1/2 years and then sold it." Following her WASP service, Nadine was a Link Trainer Instructor in Portland, Oregon for a few years.

TO COMPETE IN AIR RACE—Red-haired Nadine Ramsey beautifies herself before test flight from Los Angeles airport in her stripped-down P-38 which she will pilot in the $25,000 Bendix trophy race starting August 30. The former Wasp ferry pilot will compete in the 2045-mile race and feels she's 'a cinch to be in the money.'

Dorothy Kocher

Dorothy Kocher was in WASP class 43-W-4.

Our bay was such fun. There were six of us, weren't there? Anyway, you couldn't find six more different personalities. Jennie Hrestu couldn't get up in the morning and "Hank" (Henrietta) Richmond was a dynamo and she would say "Come on, Jennie, you can do it!" I went to St. Joe (Missouri) for Instrument school first, got my instrument card and first pilot rating on the B-25, and from there went straight to pursuit school at Brownsville, Texas. From Brownsville I went straight to Orlando for a month, and then I went back to Long Beach.

I was given a P-51D model to ferry. The D's had a fuselage tank, the belly tank, that made it so heavy. I had only flown P-51As, you know the "Awesome Dragon," and I had not flown anything for a month. I was told to fly around the field for a half hour to burn the fuel out. I was always such an eager beaver; I couldn't wait to get gone. After ten minutes, I came in for a landing and just darn near scuttled the airplane. The worst landing I ever made in my life! I dropped it in about 30 feet. The tower called me to come in, they put me in the back seat of an AT-6, and I had to check out in an AT-6 again before taking off on another delivery.

But I had landed different than most people. I never knew that, but Mr. Bolen told our class that out of all of his students, only one landed like he did. And that was me. I don't know what I did. I just flew from instinct the way my old instructor taught me. I wouldn't fly down with a lot of speed; I would fly down in a stall and I always greased them in. It just mortified me! It was just that I couldn't wait to burn out that gas; I was too anxious to go.

I was such an impatient person... always anxious to get going. I just couldn't sit up there and burn up that gas as I was instructed to do. Besides, I was one of those "Know it all's," which I think a really good pilot has to be. In other words, you have to believe in yourself that you are flying the airplane and not vice-versa.

★★★

Let me tell you of a little incident that happened at St. Joe. As you probably did, we had to all go before a board of Army Officers before being assigned to either B-25s or A-20s.

Anyway, when I went up to the Board, one of them said, "Since you have been successful politically at Long Beach, we will see that it continues, and you can have B-25 time." So help me, I did not have the vaguest notion what this Army Officer was talking about. You can ask any of the people I flew with and they will testify that I never ever did use any means to get ahead like that. It was just that I was so in love with flying and the fighters that I was like a person high on drugs all the time!

I did hear rumors that one WASP at St. Joe used her friendship to have her assignment changed to B-25s. She was killed at St. Joe in a B-25 while I was there! Fate! A funny thing! I think she was a good pilot, too. I was asked to go home with her body; but I was so near finishing. Another WASP, who washed out and elected to resign and go home, performed that function.

★★★

I don't think we had P-38s at fighter school; I think at Long Beach. Much later? Yes. But fighter school was very easy for me.

★★★

A real freaky thing happened. I delivered a BT-13 into Wichita Falls. We'd flown BT-13s only for a long time. We hadn't flown anything like an AT-6 from

First Lt. Dorothy Kocher Olsen, who formerly operated dancing school in Portland, was only woman among 350 air force reserve officers who recently completed refresher course at Long Beach, Cal. A World War II WASP pilot, she is again in military service at Long Beach as a ferry pilot for the U. S. air force.

the time we were out of school, and this was several months later; I think Iris Cummings was with me. This Operations Officer evidently didn't go by standard operating procedure (SOP) because he asked if we would like to ferry ten A-25s. I didn't even know what they were but I said "Great, yes." So he took us out there and I thought "Well, he'll get back in the seat," you know, and he'll let me fly around the pattern. But there was no back seat and it was like a fighter. If I remember, you flew it from the front seat and the gunner sat in the back. So he showed us the cockpit and said "Okay, take it around the field" and Iris and I did, and we both delivered A-25s. And from then on, I don't know if Iris got any, but I

Douglas SBD

got to ferry a whole lot of Navy Douglas Dauntless A-24 (SBDs) from San Pedro because there weren't A-24 pilots on our field.

★★★

But the best part of it was when I went to fighter school. It was all old stuff, getting into an airplane, you know — by yourself and not having an instructor in the back. But that's a mental hurdle.

★★★

I wrote in one of my letters home to mother that Evelyn Sharp was going to be in charge of Palm Springs and she was taking Jennie and me with her. But it never happened; Evelyn was killed.

I have something somewhere that said only 24 women reached Class IV status, but a lot of the P-38 pilots probably didn't have heavy Class IV time; but the P-38 wasn't class I, II or III. We were rated Class-III-P if we had qualified in fighters. We had to have twin-engine, medium-bomber time to get into P-38s. But it could have been DC-3 and that wasn't class IV, just B-25s and up.

I had a night check in a P-38 and I ferried them at night. At the last. (Kay: I think you're the only one who flew at night.) I flew into Love Field at 2:00 a.m. one morning. The most beautiful night I've ever spent in my life — bright Texas moon, 2:00 in the morning.

The last P-38 I ferried was a short trip in California, and Rex Mays, the famous car racer, came and picked me up. He was in transition in Long Beach.

The Long Beach base gave each WASP at the party a picture of herself and her favorite airplane. I have a P-38 photo autographed on the back "The gal who delivered all our airplanes, Lots of Luck." They also autographed the back "To our P-38 pilot, we will miss you." "Best wishes to a HOT pilot. I've never had the pleasure, but I do believe the boys."

In one of these letters I wrote home to my mother that I flew nine hours one day in a B-25 — three students, 3 hours each.

★★★

I got to ferry two A-20s to Montreal, Canada. In one of them, I had a hitchhiker, and I told the young man "This is the first time I have ever flown this airplane, so you ride at your own risk." He was gung-ho to go. He said "I'm not the least bit afraid." After two or three stops, he said "Everytime we stop, the men ask me if that gal can really fly that airplane?" I wonder where he is today. I wonder.

I went into P-63s on October 7, 1944 and I ferried one right after that. The reason I did that was because they would keep my P-38 in transition until it was too late for me to take off. Since I wanted to take off the same day, I got started on P-63s. It was so much fun, I just kept on. My log shows I flew a P-63 on October 2, 1944, but I'm sure that it was before that. Though maybe not, because we were flying P-39s. P-63s came at the last. We first ferried P-39s on April 11, 1944 while we were in school.

★★★

I wrote often to Mother; I wrote about flying, not about personalities. It was the airplanes I thought about.

★★★

When I was in class at Sweetwater, almost everyone in my flight class washed out. We had that instructor McKinney, remember? Almost every one of his students washed out. I wasn't exactly having trouble, but I wasn't doing good. One day I traded places and jobs with my instructor to make him feel just what it was like to be a student again! He was completely demolished!

While training in Houston, we lived in a motel where the cockroaches were bigger than we were. Jean Landis, Jennie Hresta, Barbara Ward and I roomed together.

Class 4 was double the size of the other classes, I think almost double (112 graduated out of 150 entered). Everyone of my flight, Flight A, flew the PTs from Houston when the school was moved to Sweetwater. Half of the class reported to Sweetwater, Flight B.

✯✯✯

When I was weathered-in away from home base, the first thing that I would do would be to go out and buy the most expensive bar of soap I could find, then I would go back to the hotel and soak. I'd also write letters, fix my nails, nothing special.

I still have a sort of THING about soap and I think it comes from being born on a farm during the depression days when many times we made our own soap and it really stinks! So even to this day, when I have a whole new bar of fresh, good soap well, I get sort of a thrill. Another thing, at home we all bathed in a little washtub, Saturday night, and the youngest was last, and I was the youngest. And most of the time, I think the bar of soap was pretty well worn down to nothing. So having a nice fresh, new bar of soap lifts my spirits. And I wouldn't part with those memories for anything. I think it would be such a shame to be born wealthy and miss all of that.

✯✯✯

When I was flying, I loved stopping at a different place every night! And I loved taking off and landing. Rex Mays, who was

Dorothy Kocher on the wing of a P-38. (Dorothy Kocher collection)

Ferrying pilots pause for a photo at the Bell factory. WASP are Dorothy Kocher (on wing), Ginny Hill and Katy Loft. The men are unidentified. (Virginia Hill collection)

an auto racer at Cincinnati, told me that every take off and every landing is a potential accident. He would take off from Long Beach and fly all the way to Love Field (Dallas) without stopping, and I would stop at every single chance I got — just to be able to take off and land because that was the only fun. I think he thought I was crazy!

There wasn't a stop I didn't like. I always went as far as I thought I could go without getting overly tired or pushing daylight. I put in long days. I didn't have fatigue; I had boredom. Don't ask me how I dealt with it, because I would have been in bad trouble. I had fun!

I loved being alone in pursuits. We were always alone on trips; we never had to fly in flights. I think even from the beginning, even when we were delivering BT-13s, we all straggled off — one would go one way, one would go another — even when we delivered at the same place.

I loved being alone; that was the thing I liked. I would not have liked flying in flights. I don't remember ever being afraid though I think there were times when I should have been afraid. I was too busy getting myself out of the situation. Really, when there's a situation, you have an emergency and you really don't have time to think about being afraid. You're busy.

One time when I was lost in a P-38 — such a stupid stunt, see, I was afraid of being caught, such a stupid stunt — I hadn't gassed up because I only had to fly from Coolidge, Arizona to Love Field, Texas and deliver, so I had plenty of gas. Well, I think it was a Navy pilot who was flying with me and we were flying formation, which was forbidden. I was flying on his wing and it was hard to fly with him, because they go so slow, you know. They could go off and leave us if they were climbing, but straight-and-level flight a P-51 or P-38 had a hard time staying with them. So I was flying on his wing for a long time, and not paying any attention — navigating was never my long suit, anyway. I can get lost in a hotel; I'll be sure to go the wrong way. As long as there's not open sky over me, I just have a terrible sense of direction.

But anyway, when I finally decided it was time to split up and fly in to Love Field, I hadn't the vaguest idea where he was, totally lost; and here I was, my gas tank was almost empty and I was in a P-38. So I headed for the nearest railroad depot to find out where I was (get my bearings); and I was **way** south. In the meantime, Rex Mays was on a trip with some of the other boys from Long Beach and they were just getting ready to go out and look for me because I was overdue. As I started back for

Love Field, my gauge got down to the EMPTY bar, and **then** I was scared! I didn't want a forced landing. So I landed at a field that wasn't too far from Love Field, but I didn't want to go on in. I just didn't want to push it. The first person who came out to meet me was a navigation instructor from Sweetwater! It doesn't seem possible. When I think back I can't believe that happened, but it did. He was teaching cadets at this field. And, of course, they were all thrilled to death. I think it was an AT-6 field. I don't even know the name of it. But it was so terribly humiliating. Being humiliated bothered me much more than anything else. Actual danger didn't bother me; it was being caught at some of the dumb things I did.

✯✯✯

Another time, I was flying a P-38, pretty near upside-down, when the nose wheel cowling door came open. It didn't come clear off, it just came loose and parts of it lodged in my wing. I was on my way in to Love Field that time also; and when I came down, the crew looked at it and laughed 'cause they knew I'd been rolling the airplane. The operations officer told me that I had dropped my gear coming in too fast. I thought that was just fine. They fixed it and I went on.

But those were lots of times I was scared — scared silly. And everytime I remember thinking I'll **never** do these things again; but I'd go right out and do it again because I couldn't live with the boredom. Flying just straight and level was just so boring. I loved taking off and landing! I should have been an instructor.

✯✯✯

I had no problem with the menstrual period. I think it is an unusual woman who had problems with her menstrual period. I was never restricted because of it. "Have you ever been flying upside down when you had the curse? Bubbles!"

✯✯✯

I feel that in my entire life the one thing that I was good at was flying.

When I was flying I didn't feel that I was tops in flying at all, though I've been told that "Pete" Madison says I was. I was very surprised after this was all over that so few people ever reached Class IV, because I never really worked for that. I just did what came naturally. I just loved it so, and I wanted to fly every single minute.

I've never thought about whether it was my own ability or whether the base gave me an opportunity. It was just things came as they were supposed to come. I don't even know who chose us to go to Fighter School. BJ Erickson would know. She should know. But I would think the people like our Operations Officers, also Transition Officers, and probably the men who cut the orders can tell which people really worked and which ones flew when they had to, because there was a difference - a vast difference. Yes. I don't think it was only ability, I think we used the word eager; but actually you enjoy your work or don't enjoy it. I enjoyed this work.

You get ahead when you're eager — when you like it. You can't fool people. You had to love it, and be doing it all the time, and not let anything else get in the way. Do you know I had the chance to go to Disneyland and meet Walt Disney? But I wouldn't do it because I was afraid that I would miss a flight out? When I think back at the things I passed up ... do you know that the girls had to drag me over to see Niagara Falls? I didn't want to see it; I just wanted to wait and get my airplane and go. I didn't want to see Niagara Falls for the first time. I think that we, the people who really loved it, just had sort of tunnel vision so that we only could see one thing, and that was getting in that airplane and getting it away.

✯✯✯

How we were chosen, I believe, was more on our flight records and our ability more than anything else. I know for sure that if we washed out of instruments at St. Joe School (St. Joseph Missouri) you had your choice of being transferred to the Training Command (which to a ferry pilot was a fate worse than death) or resigning and going home, which some did. So since it was necessary to get through St. Joe to be sent to fighter school, that was one criteria used. I don't think it was based on hours. I really think it was based on our performance in flying, being up-graded in transition.

I never knew how they chose those of us who were sent to Orlando for officer's training. I was sent directly from Brownsville Pursuit School, and was out of the air for over a month. That really bothered me, because I wanted to be flying. I thought it was such a waste of time.

But I think somebody had in mind that we might be militarized, and then it wouldn't have been a waste. I was assigned to Long Beach.

★★★

The only time I was ever lonely was on Christmas Eve; but even then it wasn't too bad. I went to Midnight Mass. I wrote letters all the time, and I got letters. I wrote all those letters home. But gee, I never felt tired.

★★★

The first time I saw a P-51, I had just one thing in my mind — to fly it! I was never that way about the P-38s; but I loved the P-38s when I got into them. However, I thought the P-51 was, and still think is, the most beautiful airplane ever made; and they'll **never** make another one that equals that.

There was one of those days that you just get off, and it was so beautiful and everything is so perfect that you couldn't ask for more. I got in this P-51, and as far as I was concerned, I was alone in the world with this -51. And I sat on the end of the runway and I got to thinking: I wonder what would happen if I held my foot on the brakes and pulled the stick all the way back and opened it up and then let it go? You know, we **always** eased forward.

I decided to do it. I went zooming out across the field, and held it down; and when it was really racing, I pulled it — I don't say more than that — off the ground and chucked up the gear and I shot up in the air. I made a beautiful loop over the field, and all the papers and the dirt from the bottom of the plane flew up in my face. I straightened out, and I didn't have the vaguest idea which direction I was headed in. But that take-off was such a high for me because the plane was such a beauty and performing like a dream. The tower called my number, and I was scared. These were the times I was scared thinking, "Oh, now I've done it again!" And he said "Come back soon!"

I got a letter from a young cadet who was at the field there, watching. He said, "I have never seen such vitality!" This compliment was the high point of my life. I thought that was nice.

I think if anybody said I was attractive that had to be that I was doing what I loved to do, and I was so happy. It was certainly not looks.

Another P-51 that I delivered to Patterson Field was throwing oil all the way from Long Beach. At each stop it was harder to land; the mechanics checked and couldn't find anything wrong. At Patterson, I couldn't see a thing in front; the windshield was covered with oil. They found the block was cracked. As far as I can remember, that was the only really serious problem I ever had with an airplane.

★★★

I was always a great one for rinsing out little dabs of wash and hanging up in the bathroom.

★★★

I loved the life we led. That was the charm of the whole thing. The suspense, never knowing what the next day would bring. The excitement. Excitement. That's the best word in the whole dictionary.

I can promise you I never abused a plane because I was so aware that a boy's life could hang on that airplane and how it performed. I never pushed them above the red-line. Up to it, yes — not above.

I had the gear hang up on an A-24 one time — one down, one up; but I got them both down and came in for a landing with no problem. I think that was Phoenix.

I remember my first landing on a snow-covered field taking a P-39 or a P-63 to Great Falls, Montana. The field had just opened, if I remember right, and I was one of the first planes in to it. The field had deep snow all over, but the runway was packed with snow on each side, piled up high. What a thrill! I had no idea how the plane would behave on a landing on snow, but it was no problem at all. What great fun!

All those things added up to make it such a wonderful experience... every day was different. When I think of those days, it's a blank in between flying.

The only special time I remember about being weathered in, I think it was in Midland, there was a boy from my hometown whose name was Shorty Wooster. He was ferrying P-51s also, and he was a practical joker. I got a call supposedly from Operations at Midland asking me why I wasn't with the rest of the planes off the field since those planes were in the air; the weather was good enough I should be in the air. But it wasn't Operations. I had been sleeping in because we knew for sure the weather wasn't going to clear. It scared the heart out of me. Shorty was just playing a big joke on me. That was the only time I remember. Most of being weathered in is just a blank.

★★★

I remember the only time I got to see New York was between delivering the airplane and going over to Niagara Falls. I don't remember the circumstances, but I do remember I had something wrong. I had a cold that settled in my eye and I couldn't keep my eye open; it just watered. It would hurt to open my eye. I didn't see Porgy and

The fuselage of an Air Transport Command plane is a backdrop for 6th Ferrying Group WASP Nadine Ramsey, Betty Tackaberry, Katy Loft, Evelyn Trammell, Thelma Harris, Debbie Truax, Ginny Hill, Carole Fillmore and BJ Erickson are standing. In front are Billie Wilkes, Lauretta Beaty, Iris Cummings, Lew Coleman, Dottie Webb, Jean Landis, Dorothy Kocher, Ruth Thompson and Helen Richards.

Bess, but I did see the Rockettes. There was someone with me, but I don't remember who. I remember going to see "The Good Earth" three times; but I never could hear the dialog.

★★★

We were treated like visiting royalty by the men we encountered. I don't know of any instance, ever, either on the base or off the base, when we weren't treated with respect. The men went out of their way. The men we flew with treated us like little sisters. There were never any passes. I've heard, or read, things that are just hard for me to believe happened. I think somebody is using their imagination and making things up. It certainly wasn't like that at our base. It was certainly out of place in those days. It just wasn't necessary, we didn't need it. We had enough excitement with our planes.

★★★

My family was wonderful. They worried about me, but they were proud of me. And I felt that I was doing something that was worthwhile, doing what I loved and doing something that counted. That was the satisfaction in doing it.

I never really thought about the fact that pursuit flying made me a unique WASP, except that I felt sorry for everybody who wasn't in my shoes. I used to be flying along in P-38 or a P-51 and looking down on all the guys and thinking, "Oh, you poor people, I feel sorry for you." I just felt sorry for the whole world except for myself. I just wouldn't have traded places with any movie star, or with **anybody** in the whole world. And I wouldn't pass it up for anything. I think it was the most outstanding experience of my whole life; and I've had a lot of unusual experiences.

Madge Rutherford

The following are excerpts of letters from Madge Rutherford to her parents.

Madge A. Rutherford, WASP Class 43-W-4.

Long Beach Army Air Field, May 11, 1944.

In case anyone wants to know if I need anything — I could use a Derringer with some ammunition. As I will be flying hot planes in the not too distant future and will carry a gun by orders, I'd prefer to have my own. These army-issue revolvers are heavy! Something small and compact would be appreciated.

Brownsville, Texas Pursuit School, June 2, 1944.

The puffed wheat candy has been wonderful. Thank you so very much. The new class has arrived and I have my ex-roommate from Sweetwater with me — A.J. May.

June first at 10:00 a.m. saw me streaking along the Gulf Coast over the ex-pirate haven Port Isabel in a P-40, all by myself and hysterically wondering if it would be better to try to land or climb to ten

Madge Rutherford on a P-39 wing.

thousand and bail out. Frankly, my inclinations were toward bailing. Oddly enough, when practicing stalls with gear and flaps down, I discovered that she doesn't stall until 75 mph. So today I am an old veteran of five flights and gaily jolt in over the boundary at 100 per, flare prettily and skip on landing. I made a couple of terrible landings today, and only a pair of well-developed legs saved me and the ship from ground-looping. Fortunately, the instructors in the little jeep by the runway's brink have chosen to wink at this, and have made no remarks about a check ride. These are given on the slightest provocation, I'm told. If I got by today, I doubt if I need worry.

My, aren't I getting blase, calmly writing about soloing a P-40? But it really isn't hard to fly, just a little more power. Had an Allison engine out in front and looking down the nose is about like driving a Duesenberg. Sherm's* not happy about my flying pursuit, and has expressed himself in no uncertain terms. I wish he wouldn't worry so about me. I'm overcautious in the air; always have been. You must not worry either.
(*Author's note: Sherm is Madge's husband.)

Really expected to fly the P-47 first, but the cards didn't happen to fall that way. The P-40 first is a break for it lands like an AT-6, only trickier. Its favorite trick is to drop the left wing and ground-loop to the right. But so far, so good.

My last flight today, and I was very tired, was mercifully short. I had made one landing and was batting about to take off when a jeep came roaring across after me, and excited flight instructors pointed out to me that my beautiful 100 octane was pouring out of the left gas vent under the wing. One yelled "Take her off and she'll stop!" Another called: "Cut her!" I did just that and got out mighty fast. These learned gentlemen could not imagine why this was happening. Meantime the 100 octane spread a lake under the empennage, and so I left them. I'll be anxious to know their decision. The P-47s siphon gas, but the P-40s aren't supposed to.

The beauty of flying liquid-cooled engines is that the coolant heats up so fast on the ground that you have to take off before you have time to reconsider. If you had time, you'd probably never take off at all. Incidentally, our ships have charming names emblazoned on their streamlined noses. I flew "The Happy Moron" and the "Mad Russian" this morning.

Time for sleep now.
Love, Madge

Sunday, June 11, 1944

We're flying today because our flight is behind in time. I've just returned from 1-1/2 hours airwork and five landings in the big old Thunderbolt, and my coordination for writing is pretty rocky. If certain cumulus over the practice area would dissipate like good fellows, I will fly my first P-39 in about an hour. But things don't look too good just now. My landings were good today, but who could miss with a P-47's 15'7" gear?

Received a card from Sherm from Honolulu. He's on his way back to the war zone now, and I just have a feeling it's on the Philippine campaign. Just a hunch, however.

Yesterday the post here staged a big war bond parade and I drew the Colonel's car to drive. Naturally, it rained. It almost never rains down here, but yesterday afternoon it poured. I was the only WASP to stay dry since the other gals were riding in Jeeps and Re-Con cars. I guess I felt a little too jubilant about it, for retribution came. I drove the car by myself back to the base. On the way, I bought an enormous watermelon. Pulling up in front of the B.O.Q, I stopped, got out, embraced the melon, took two steps and sprawled, soaking me and uniform and spreading watermelon over a radius of ten feet. I didn't get hurt, but I am going to have to practice wearing hi-heels again. "Pride goeth before a fall and a haughty spirit before destruction."

Perhaps I may fly that Bell Booby Trap yet. I guess I'll go and try to convince the dispatcher to give me #19 which is Queen of the 39s here.

Dad's letters have been swell. Sorry you've not received word of me for so long. I'll try to do better.

Love, Madge

Dallas, Texas, June 17, 1944

Yesterday noon we and luggage boarded a C-47 and bade Brownsville farewell. We had quite an impressive graduation ceremony complete with parade and were not a little tired. At 3:30 we arrived here in Dallas and remained overnight awaiting orders. Theory is, that Kitty (Leaming) and I, the only two from HB (Long Beach) will get P-51 checks sometime today and then be ordered to Newark. But that remains to be seen.

Fortunately, the candy and cookies arrived before I left and they are wonderful. I send you greetings from various half-starved ferry pilots who agree that you are a wonderful cook.

I hold ratings on P-39s, P-47s and P-40s now, and the P-47 is my baby.

Chief worry at present is this blasted bill and today's D-Day on said measure to deactivate the WASP. I sent Ramspeck another wire last night but I guess all we can do is wait and see.

Although not definitely decided, I think I can say you may expect me home if the measure is passed.

We had quite a nice farewell party (Brownsville) and wept on one another's shoulders. But I'm sure it wasn't for the airplanes, for the ships we flew were war-weary and haggard and were prone to make their pilots likewise. I came off very clean; having only two emergencies: one in a P-40 when the gear and flap system went phoey, and one in a P-39 when the door sprang open at 9000 feet. Fortunately, I was able to deal with these little matters in the approved manner; but let me tell you, I really put my foot down about flying P-39 #10 again, and put it on a Red X until the springs are fixed. That's one of the beauties of being a peashooter pilot, you really have the final say on the ships you fly.

Guess I'll go check again on the P-51 deal.

Happy Father's Day, Dad. Maybe you'll find some use for that knife.

Love, Madge

Chicago-In Flight Offices, June 21, 1944

Dear Mother and Dad: Arrived last night in Newark and stayed over to see if I was needed to ferry a P-63 to Great Falls. The Major decided that P-51s are more important, so I'm on my way to California and am about to land in "Chi" (Chicago). I flew over 1600 miles yesterday and am happy to ride for a change. Upon refueling in Greensboro, S.C., I learned that Headquarters had been wiring ferry bases all over the country to "Get the 51s to Newark!" Seems they're vital. So it's back to Mines Field in L.A., for another Mustang.

Love, Madge

Evelyn Sharp

From Ord, Nebraska, Evelyn was one of the original WAFS. She had soloed at the age of 14 in the days when it was almost unheard of for a girl to solo at all. Local business people of her hometown, realizing her unusual potential and the need for pilots, decided to raise funds for Evelyn's own plane. She got her transport license May 13, 1938, and was acclaimed as the youngest female pilot in the United States with a commercial rating. Evelyn flew in Ord's first airmail letters, and the whole town turned out to see her come in.

Shortly after World War II broke out she joined the Women's Auxiliary Ferrying Squadron. She ferried different types of planes and was killed when the twin-engine P-38 which she was flying had one of its engines fail on take off near New Cumberland, Pennsylvania, on April 3, 1944.

A lasting memorial is laid out on the runways of Evelyn's hometown in the airport which bears her name. It is called Sharp Field. There is a plaque and museum at the airport. A local businessman wrote, "If you knew how many people involved in aviation helped out in this project, you would be amazed."

Evelyn Sharp gives the "OK" from the cockpit prior to take-off. (Barbara Erickson collection)

Ginny Hill

I graduated in August, 1943, WASP class of 43-W-4 from Sweetwater. We were that half of the class that started in Houston, then got sent up to Sweetwater to finish out Basic and Advanced. I already had a commercial license and an Instructor's rating when I joined up. Thank goodness, or I'd never survive some of the lousy instruction I got from some of the misfits that were hired to teach us. (I had some good ones, too, though.) Because the Houston branch of 43-W-4 was kept in one flight, we never saw much of, or got to know much of, the other half of the class that had started at Sweetwater. It was like another class.

Long Beach, the 6th Ferrying Group, was a great group, and BJ Erickson was in charge of the WASP Squadron. Long Beach was a much happier place than some other bases, probably due to BJ Erickson! I remember, or experienced, little of the petty jealousies and bitching.

Seems to me I was always getting shoved into transition for a hotter ship sooner than I really wanted to. I went to Instrument School in St. Joseph, Missouri, in April and pursuit school in Brownsville in May. My "diploma" on June 1st qualified me in P-47s, P-40s, P-51s and P-39s.

On my birthday, October 24, 1944, I got checked out in an A-20, which gave me a Class IV rating. Also got checked in B-25 as co-pilot but never delivered either airplane. I did deliver a B-17 as co-pilot for BJ Erickson. I got put in transition for twin-engine pursuits. After hanging around the base for two weeks waiting for a P-38 to get my time in so I could get back to ferrying, I checked out in the P-61 Black Widow Night Fighter, and delivered several of them to the East coast. Then I finally got checked out in the P-38 and delivered several of them before we were disbanded. Most of my ferrying career was in P-51s and P-63s. I remember having been at the controls of ten different planes before I duplicated any.

I was lucky to be based at Long Beach where so many planes were made.

A P-61 tail serves as a backdrop for Virginia Hill. (Virginia Hill collection)

These usually had to go to the east coast.

We never did have to travel in "flights" like the WASP at Wilmington. Just before I got to Long Beach, some flight leader had led his flight of a dozen P-51s into Mexico — he forgot to go East — and the planes were confiscated for several months. After that, Long Beach just gave individual orders and we were on our own to deliver them to wherever the orders read.

Oh, yes. I got sent to Orlando Officer's Training School, for whatever reason the Army only knows. That was in August 1944. At least none of the women pilots outranked anyone else, since we remained civilians. I am sort of glad we didn't have rank, with all the petty jealousy and paranoia. We all got the same pay.

★★★

I still have a trunk full of my orders and the check list for every plane I flew.

P-61 Northrup Black Widow Night Fighter has two 2000 Horsepower Pratt and Whitney engines. Due to the complexity of production they only saw one year of action, reaching the combat zones in Europe and the Pacific June 1944. There were only 674 built before the end of the war with Japan, and 32 of these remained in use until 1950. (David Sams collection)

I got checked out in a P-61 because I was tired of sitting around the base grounded until I got my twin-engine pursuit rating, and all the trainer P-38s were mechanically out. The P-61 gave me the rating, and all I needed was four hours solo in a P-38 and I was off ferrying again instead of sitting around the base. There were at least five or six other WASP from Long Beach checked out in the P-38s. I think all of us were checked out in all pursuits except the P-38 and P-61 by the end of the WASP's existence at Long Beach. I don't remember my being unique in flying them then.

WASP and Ferry Pilots at Bell Factory with P-63. Jo Myers, L. Perkins, E. Fisher, D. Webb. (Enid Fisher collection)

Helen Richards

BJ: Helen Richards, Sidney Miller and I are the three from Dallas in Pursuit School.

K: Helen Richards from Boise went on to Long Beach after that, she didn't go back to Dallas. She transferred after Dorothy Scott was killed.

BJ: There were just three of us from Dallas. They told us in just two seconds. We were ferrying from Grand Prairie the shuttle, AT-6s bloop, bloop, bring them from Hensley to Love, Hensley to Love and then by bus back to Hensley. All day long.

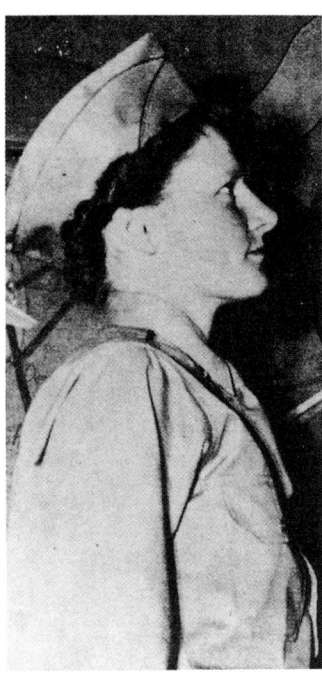

Helen Richards WAFS in 6th Ferrying Group, Long Beach.

Lana Cusack is ready to ferry a twin-engine Beechcraft. Note her issued flight suit with billed cap in left hand, knee pockets for maps and tape above right knee to steady the in-flight log. (Collection of Martha Boxberger, Lana's daughter)

Lauretta Beaty

Lana Cusack

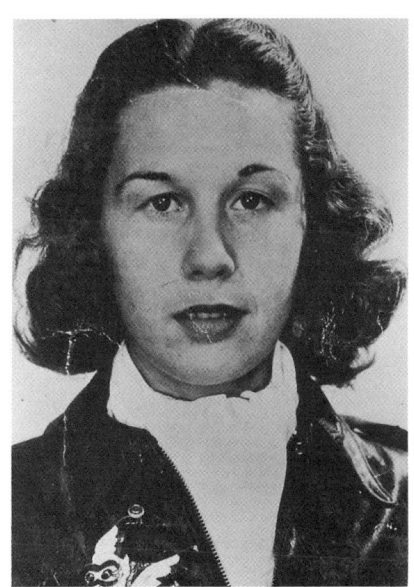

Nancy Featherhoff

Enid Fisher

Thelma Nadine Harris

Virginia Hill

Dorothy E. Kocher

Jean Landis

Kittie L. Leaming

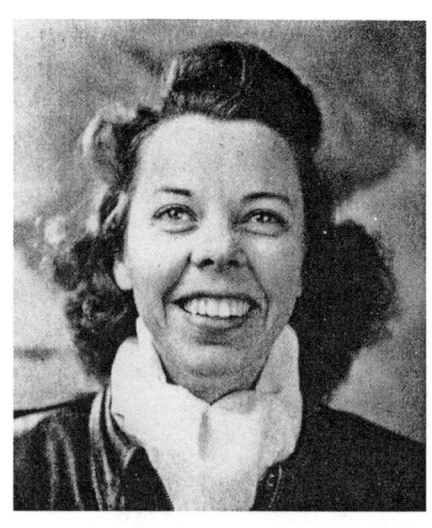
Dorothy M. Nichols

Nadine B. Ramsey

Madge A. Rutherford

Betty C. Tackaberry

Evelyn Lacy Trammell

Debbie Truax

Dorothy E. Webb

Rena D'Arcy Wilkes

Long Beach final banquet with ATC-decorated cake reading "WASP - Good Luck - Happy Landing" are WASP in finery and flowers. Top row are Ginny Hill, Iris Cummings, Lauretta Beaty, Evelyn Trammell, Dottie Webb, Dottie Kocher, Peggy Calhoun, Billie Wilkes, Katy Loft, Jean Roberston and Betty Tackaberry. Beside the table are Ellie Dressen, Elaine Holt, Ruth Thompson, Jean Landis, Barbar Erickson, Deb Truax, Lew Coleman, Carole Fillmore and Nadine Ramsey. None of the men are identified. (Rena "Billie" Wilkes collection)

On Long Beach parade grounds with Signal Hill oil wells in background are WASP BJ Erickson, Lauretta Beaty, Lew Coleman, Carole Fillmore, Ruth Thompson, Iris Cummings and several unidentified male personnel. (Lewise Coleman collection)

21st Ferrying Group

21st Ferrrying Group
Air Transport Command, Palm Springs, California

Name	WASP Class	Pursuit School	Graduation Date
Alexander, Eleanor	43-W-6	44-18A	10/1/44
Cadman, Vivian	43-W-5	44-15B	8/15/44
Castle, Margaret L.	43-W-6	44-13A	7/15/44
Colbert, Mary Lou (2nd Squadron Leader)	43-W-1	44-7	4/44
Edwards, Rebecca H.	43-W-6	44-16A	8/31/44
Franckling, Ruth "Ruthie"	43-W-2	44-8A	5/1/44
Granger, Byrd Howell (1st Squadron Leader)	43-W-1	44-5	3/8/44
Johnson, Vega	43-W-1	44-8A	5/1/44
Perkins, Lola C.	43-W-6	44-14B	8/1/44
Russell, Barbara	43-W-2	44-8B	5/1/44
Trotman, Mary Tufts	43-W-2	44-6	3/44
Vail, Catherine "Cappy"	43-W-2	44-3	1/44
Wagenseil, Martha D.	43-W-2	44-8A	5/1/44

Byrd Howell Granger

Mary Lou Colbert succeeded Byrd Granger as Squadron Leader of the 21st Ferrying Group WASP, and served until they were disbanded December 20, 1944.

Ruth Franckling

Ruth Franckling

Ruth Franckling had her Commercial License and was an Instructor with 700 hours before entering the WASP program in the second group, 43-W-2, to go through Jacqueline Cochran's school at Houston, Texas. She earned her Instrument rating, multi-engine, and completed pursuit transition to the fighters in the first group of pilots to be trained at Brownsville, Texas, when the school was moved from Palm Springs.

When Palm Springs became the 21st Ferrying Base, Ruth was stationed there. She recalls that a typical week was taking a P-51 from the West Coast factory; flying 10 hours to Newark, New Jersey, with overnight or refueling stops; hopping a flight to Niagara Falls, New York; picking up a P-39 at the Bell Factory there; flying 8-1/2 hours to Great Falls, Montana; hitching a ride to Long Beach or Palm Springs, California and starting all over again.

Only twice did Ruth experience trouble. Once, after having her P-51 mysteriously grounded for two days at Tulsa, Oklahoma; and finally given the green light for take-off, she was about to lift the gear when the engine quit. Ruth set the plane down and wore the tires down to the wheel hubs trying to stop before the plane hit the gas trucks, administration building and the watching crowd. She was recommended for a commendation for this landing but never received one. This aborted flight confirmed the fact that there was water mixed with gas in the field's underground tanks.

The other close call came while flying a P-63 Bell Kingcobra at 8,000 feet over St. Paul-Minneapolis, Minnesota, and switching gas tanks — again, a dead engine! Ruth dived to get the windmilling prop to start the engine; it coughed and started, and Ruth was able to make it to Fargo, North Dakota. She discovered the Niagara Falls Factory crew had forgotten to remove plugs in the wing tanks which would have allowed the gas to flow freely.

In spite of these few hectic moments, Ruth's skill as a pilot overcame these problems. She recalls that she loved every bit of her assignments, and was bitterly disappointed when the program ended December 20, 1944.

"We offered to stay on a $1.00 a year, but they wouldn't go for it."

Bell P-63 flies above a P-59 jet.

Catherine "Cappy" Vail

"Cappy" Vail had a total time in the WASP of 34 hours and ten minutes in flying the P-38s. This was an amazing amount of flight time considering she had her first check out in the P-38 on October 26, 1944 and the WASP were dismissed and could no longer fly after December 20, 1944.

✯✯✯

"I was at the 5th Ferrying Group, Love Field, Dallas, Texas, and went to Pursuit School at Palm Springs, California, in December 1943. Betty Whitlow, Jane Emerson, Sidney Miller and Betty Bachman were in the class with me. My first pursuit plane flight was on December 17, 1943, a P-47. From then on, in the WASP, I flew pursuits, fighters, practically all the time until we were disbanded. My first P-38 flight was October 26, 1944 in Palm Springs, preceded by a check out in the A-20. I had transferred from Dallas to Palm Springs a few months earlier. Looking back, I realize I was worried about Art. We had been married in June, and he was off fighting the Japanese, based on Iwo Jima and flying fighters. I had such a wonderful time in pursuit school in December. It was closer to my California background, family and friends. During the last months that we served, I think now that I was somewhat demoralized. However, we did what we had to do without feeling sorry for ourselves.

"I wrote Art extensively about my experience in 'fighters.' He was flying

Cappy Vail models Houston trainee flight suit.

Cappy Vail, hot-shot pea-shooter pilot, stands beside her P-38. (Cappy Vail collection)

them, too. That is why I always liked the P-51 most of all. Unfortunately, in his moving around the Pacific, he wasn't able to keep my letters. However, I did write my mother and father, somewhat expurgated so they wouldn't worry too much! And those letters bring back memories. At first, I could remember every flight, based on my logbook; but it took me over a year to get over being discharged before the war was over.

"Yes, I was in the last day picture at Palm Springs. I had just brought in a P-51, my last flight as a WASP, and I was met with the news that I must appear in a uniform at the flight line for closing ceremonies. There wasn't even time for a shower! Yes, I remember the occasion well. It was an upsetting time."

★★★

Cappy Vail fell of a wing! Or so the story went.

Cappy Vail maneuvers on crutches (Yearbook)

But she didn't!

In training at Jacqueline Cochran's Houston school in the second group of women pilot "trainees," class of 1943 designated 43-W-2, Cappy tells of the accident:

"I ruined my leg by crashing into a door frame."

"Was it the Ready Room?"

"Was it the john?"

"Actually, I had poured water on Ellen Gery. Ellen was after me, full speed... seeking revenge."

Cappy was in deep trouble. Possibly she would miss flying time and be dropped back one class. Possibly she would be permanently dropped from the program. Injured pilots were not tolerated. But this great gal was too precious to leave behind or let drop out of the program.

She could still fly!

Her classmates didn't even consider her graduating with the next class. So these classmates hauled her crutches and parachute back and forth to the plane so that she could keep up. She did, and she flew on to become a valued fighter pilot, eventually, as Class IV Pursuit, Instrument-rated, and flew a P-38 with the Ferrying Command.

She says: "What's a little accident when you've got friends like these?"

★★★

December 10, 1944, just ten days before the WASP were deactivated and could no longer ferry aircraft, Cappy Vail and Kay Gott took a B-25J, North American Mitchell bomber from Cheyenne, Wyoming, to Great Falls, Montana, dividing the time as first pilot and co-pilot so that each woman got one-hour-fifty-minutes' first-pilot time. To do this, and each have one landing, they refueled half-way, and changed seats at the stop. They flipped a coin to see who should fly the first leg.

In Cheyenne there was an Air Force "non-com" serviceman waiting to hitch a ride. Kay and Cappy agreed with Operations to take him along, even though there were regulations against "unauthorized" passengers. Cappy and Kay had both been grateful themselves to grab a ride home. So

"North American B-25. This 14-ton giant has speed and maneuverability which the enemy has yet to equal." (NAA photo postcard)

P-38

when he arrived at the bomber on the line to learn who would fly it, his consternation at flying with two women pilots was overcome by his urgency to get home. The poor fellow never relaxed the whole trip!

Kay had delivered a P-38L on the 9th of December to Great Falls, and Cappy had delivered a Bell P-63. As Kay was based in Dallas, and Cappy with the 21st Ferrying group at Palm Springs, this timing to meet together at Great Falls, Montana was most unusual. Cappy and Kay were classmates in Houston WASP training class 43-W-2, and friends in Dallas at 5th Ferrying Base. Cappy transferred to Palm Springs 21st Ferrying Group about June or July of 1944. Due to the nature of the ferrying job, it is extremely unlikely for paths to cross, much less join up. So this B-25 trip together was a very unlikely event; just a strange coincidence that two friends, two women highly trained in flying, both qualified Class IV Pursuit pilots, both with Instrument ratings, both ferrying fighters should arrive at the same time, get another set of orders from a third base, Great Falls, to fly together and to deliver this B-25 to Great Falls. A memorable trip for both women!

Vivian Cadman emerges from the cockpit of a P-63. (Vivian Cadman collection)

Vega Johnson holds onto air intake scoop of a P-40. (Vega Johnson collection)

Vivian Cadman

Vivian graduated September 1943 as a member of WASP Class 43-W-5.

★★★

"In the fall of 1940 at Fullerton Junior College, California, I talked my way into a CPT class of 48 boys. The program director really didn't want 'a girl' in his class. However, in the class that followed mine, there were two women who were accepted. One was Betty Naffz, WASP class 43-W-4, who just happened to be visiting her brother in California and took up flying for the lark. We didn't meet then because I had transferred to the University of California, Berkeley.

After Sweetwater, I ferried out of Love Field at Dallas, Texas until the Palm Springs WASP Squadron was formed. The Palm Springs WASP squadron was unique, I believe, in that all of us women were fighter pilots — pursuit pilots, as the Army Air Force termed it then.

Margaret Castle and Nelle Carmody pose in their flight suits. (Margaret Castle collection)

Mary Lou Colbert and Byrd Granger were the Squadron leaders. I was in Dallas with the 5th Ferrying Group when they asked me if I would like to go to Palm Springs with them to fly pursuits . . . "Yes!"

I received my Instrument White Card rating at Palm Springs, June 17, 1944. Took my rating on B-26, and delivered one C-47 as first pilot. Then out of Palm Springs I went to Brownsville, Texas for pursuit school, August 15, 1944 to September 15. Then back to Palm Springs for duty.

A routine trip for us was to pick up a brand new P-51 Mustang from the North American plant near Los Angeles, fly it to Newark, New Jersey, then take the train up to Niagara Falls to pick up a Bell P-63 to deliver to Great Falls, Montana. At Great Falls they would load us into a C-47 and we would be taken back to Palm Springs. After a night's sleep and a change of clothes, we would do the same thing all over again. A "routine trip for us" was not our "daily" routine! I remember many a day of boredom sitting around waiting for orders.

★★★

In my mind there is such a vivid picture of being there when Hazel Ying Lee was killed. Of being in Operations at Great Falls when we heard the horrible story of another plane landing on top of Hazel.

My flight time shows three hours fifteen minutes in a P-51D on November 23,

North American B-25A "Mitchell" Aircraft.

that fateful day. On November 24, my records show one hour fifteen minutes in a P-63.

★★★

It is still hard for me to believe that we were allowed to fly the hottest stuff the Air Force had then. I checked out in every Pursuit except the P-38. That transition in the 38, and Officer's Training School (OTS) at Orlando were to be my next assignments. I received Class IV rating, but declined the opportunity to fly the P-38 the last months before deactivation December 20, 1944.

In the last days photo of the WASP formation at Palm Springs, you can see that we are all clutching our walking papers tightly in hand. What an honor to have been given such a privilege! And given a chance, I'd do it all over again!

Lockheed P-38H "Lightning"

Martha Wagenseil climbing out of cockpit. (Martha Wagenseil collection)

Eleanor Alexander   Vivian C. Cadman   Margaret Castle

Mary Lou Colbert   Rebecca H. Edwards   Vega Johnson

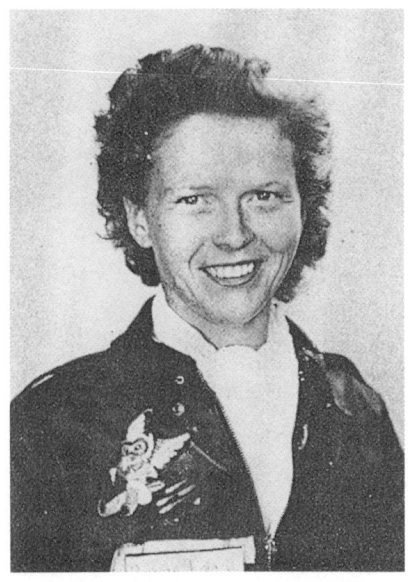

Lola Perkins

Barbara Russell

Mary Trotman

Martha Wagenseil

21st Ferrying Group WASP last day, Dec. 20, 1944. Mary Lou Colbert, Martha Wagenseil, Margaret Castle, Catherine Vail, Ruth Franckling, Vega Johnson, Vivian Cadman, Lola Perkins, Becky Edwards, Byrd Granger.

Epilogue

Recognition of the WASP — Women Airforce Service Pilots — as World War II veterans only came about in 1979, some 37 years late. These women pilots who served over two years under Civil Service, 1942-1944, were awarded military status by an act of Congress.

★★★

Women today should know that women pilots accomplished this task and did it well. That the WASP were unfairly dismissed before the war concluded does not diminish the fact that women pilots served the country in its time of greatest need, and did it with distinction.

This is true of all the WASP who flew in World War II, not just for those few in this document who served in the Ferrying Division for the Air Transport Command, ferrying fighter airplanes.

★★★

Kay Gott, 43-W-2, ferrying pursuit out of the 5th Ferrying Group, Dallas, Texas, reminisces with Betty Shea, 43-W-5, ferrying pursuit out of the 3rd Ferrying Group, Romulus, Michigan.

Kay: It will be nice to know our own history - B-26 story, the remote control PQ-8 and PQ-14 techniques developed then, and used in 1991 Desert Storm targeting.

Betty: There are three stories I would like to know - Camp Davis, the ATA Story in Britain, and **why** we went home before the war was over.

Kay: I want to know the tow target story, not just Camp Davis. What about the girls who flew targets over the Oregon desert - that would be a fascinating story.

And down at Yuma, Arizona where it was so terribly hot - what an epic. Then there are the gals who taught Instruments at Clovis, New Mexico - that would be an interesting bit - and did you know we had some Instructor Pilots WASP at Randolph Field, Texas. I found that out in 1986 when Women Military Aviators met at San Antonio, Texas. Here was a photograph of WASP at Randolph. I asked "What in the world were they doing here?"

We don't even know our own history. It's the pits!

★★★

The greatest assembly of women pilots of achievement gathered for the Women Military Aviator's (WMA) conference October 1991 in Sacramento. Present were members of the women who flew for the government in World War II - the Women Airforce Service Pilots. Most exciting were the women "firsts" in aviation; women who served in Desert Storm in January 1991; Lt. Col. Kelly Hamilton, who has been WMA president for the past four years; Commander Rosemary Mariner, one of the first eight women to enter the Navy's military flight training, earning her wings in 1974; Mrs. Jean Kaye Tinsley in helicopter "firsts"; Jacqueline S. Parker, Captain, USAF, "first" Experimental Test Pilot, now at Wright-Patterson AFB; Major Eileen Collins, NASA candidate to be the first woman to pilot command a space shuttle; Stephanie Wells, NASA Research Pilot. These women are among the "greats" in Women in Aviation making HISTORY TODAY!

★★★

So what's the fuss about women pilots in the '90s? Women pilots can't fly this and that airplane? Who says they can't? The President? Not Congress.

My heaven, what was that all about in 1944, way back in World War II, when women flew every airplane that was made then? The airplane doesn't care whether the pilot is a man or woman. It boils down to who can handle the craft. A fast airplane demands a skilled person, man or woman.

And by the end of 1944, women proved they could handle anything flying then.

Let women fly, and let them train and qualify. They can do it. Our group proved it.

WASP Barbara Ward, Class of 1943-W-4 and WASP President 1988-90 wrote in the October 1989 Women Military Aviator's Newsletter that Canada, Belgium, Sweden and Netherlands employ women to fly combat aircraft. Canadian women pilots today are flying jets worth thirty-five million dollars each. Not so, U.S.A.!

At this writing, women in the United States are prohibited under Congressional law from flying about 26 airplanes, which are classified combat aircraft. The bill to allow American women pilots to fly those airplanes was in committee, to study this issue and report to the President of the United States in November 1992.

★★★

Killed in ATC Service

Name	Plane Type	Location	Date
Cornelia Fort/WAFS	BT-13A	near Merkel, TX	Mar 21, 1943
Virginia C. Moffatt/43-2	BT-15	Ontario, CA	Oct. 5, 1943
Elizabeth M. "Met" Trebing/43-4	PT-19	south of Norman, OK	Nov. 7, 1943
Dorothy F. Scott/WAFS	BC-1	Palm Springs, CA Pursuit School	Dec. 3, 1943
Evelyn G. Sharp/WAFS	P-38J	near New Cumberland, PA	April 3, 1944
Dorothy M. Nichols/43-2	P-39Q	Bismarck, ND	June 11, 1944
Susan Parker Clarke/44-2	BT-13B	Columbia, SC	July 4, 1944
Paula R. Loop/43-2	BT-13B	Mts. near Medford, OR	July 7, 1944
Alice E. Lovejoy/43-5	AT-6	Brownsville, TX Pursuit School	Sept. 13, 1944
Gertrude V. Tompkins/43-7	P-51D	left Los Angeles	Oct. 26, 1944
Hazel Ying Lee/43-4	P-63A	East Base, Great Falls, MT	Nov. 23, 1944

Remarks for WASP Memorial
9-26-86
Sweetwater, Texas

We gather here today to pause and to
remember. To remember friends, special
friends.
Friends with whom we shared the unique
sorority of the WOMEN'S AIRFORCE
SERVICE PILOTS.

As WOMEN AIRFORCE SERVICE PILOTS,
we have shared the wonders of flight:...
The spiritual uplifting of exploring God's
heavens,
the awe of heights unknown to earthlings,
the expansive beauty as seen
by eagles
and by pilots.
These intangibles have bound us together
forever.
For them, no words exist.
They are known by pilots
and can be shared
only by other pilots.
When invaders threatened our shores
during World War II, we came to the
WOMEN' AUXILIARY FERRYING
SQUADRON and the WOMEN
AIRFORCE SERVICE PILOTS program.

From protected homes
and loving families,
we came to offer ourselves,
our skills as pilots,
to our country
at war.

As in all wars,
we offered our lives.
Thirty-eight of us paid that price.
WE PAUSE TO REMEMBER
these friends...
these sisters who died during
World War II.
Since those days,
more than 40 years ago,
many of us have joined those 38
WE PAUSE TO REMEMBER
these friends...
these sister pilots who shared our skies.

There are special memories we all share:
our leader and commander
has gone.
a pilot of exceptional caliber,
a blond, beautiful woman who battled
chauvinistic bureaucracy
(and usually won).
Our Establishment Officer, confidante,
overseer
and mother superior, left us this year.
WE PAUSE TO REMEMBER her,
a woman who loved each of us,...
who suffered many frustrations for us,
as well as from her last painful years.
WE PAUSE TO REMEMBER, ...
and we weep.

But,
WE PAUSE ALSO TO REMEMBER
Our own frailties.
For we, too,
will soon fly behind the dark, towering
cumulus.
We shall join our sisters
in the sun-speckled
reaches of space,
at altitudes we never dreamed of,
in that special part of heaven
God has reserved
for his pilots.
There, our planes will be infallible.
Our engines will never fail us.
Our aerobatics will cut incredible figures
across the face of heaven.
We shall never be afraid.
And with those who have preceded us,
our joy shall be forever.

by Dora Dougherty Strother, WASP, 43-W-3

★★★

    The parents of those killed in the
line of duty paid to bring their bodies home
since they were not classified as military,
thus entitled to benefits afforded the military. Often times, a collection was taken
among the WASP to help defray the funeral
expenses.

Symbol of the Air Transport Command with Globe and Spire and a Morse Code A.F.A.T.C.
Women were Auxiliary - WAFS - before name changed to Women Airforce Service Pilots - WASP.

319TH. A.A.F.F.T.D.

Fifinella gremlin copyright by Walt Disney. Her use was granted to the WASP for their logo. She was cross-eyed and had horns, a mischevious flyer!

Women in Pursuit 215

Matchbook covers of World War II. Army Air Corps shoulder patch wings and star from WWII.
Cloth jacket patch symbol of the Air Transport Command - the Globe and Spire and in Morse Code A.F.A.T.C.

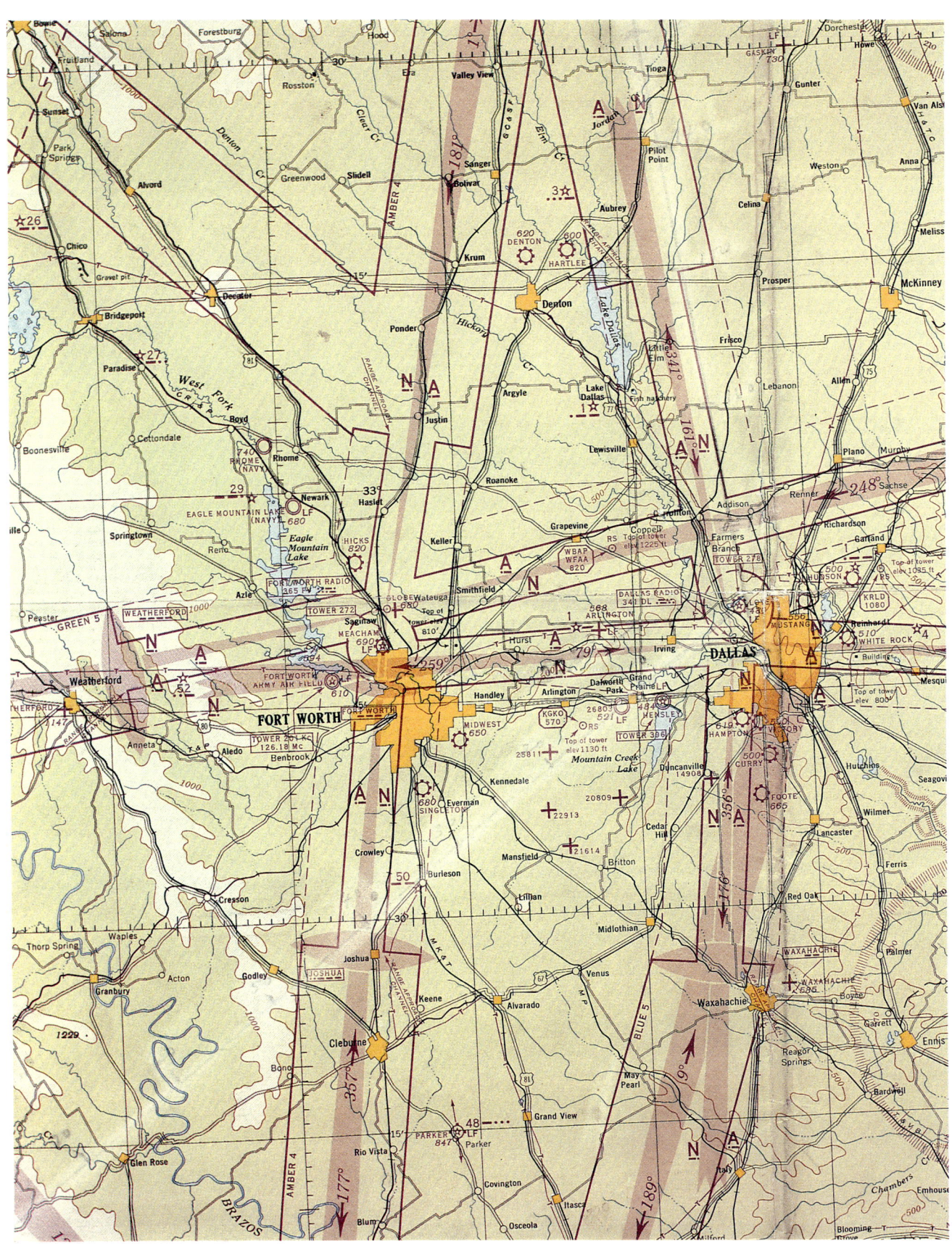

1944 Dallas flight route map.

Appendices

APPENDIX I
AIRCRAFT DESIGNATION SYMBOLS
(used in World War II by the Army Air Corps)

"**A**" designated the attack aircraft used in World War II. WASP flew A-20s and A-24s to tow target sleeves for the ground gunners to learn their craft, i.e., shooting at a fast-moving target in the air. Sometimes these planes came back with holes in the plane rather than in the target sleeve. At Camp Davis, South Carolina, WASP were used to fly these missions. Two women, Mabel Rawlinson (WASP Class 43-W-3) and Frances Grimes of the same WASP class, lost their lives in this dangerous exercise. The A-24s were single engines, and the A-20 was a twin-engine.

"**B**" was then, and still is today, the designation for the bomber airplanes.

The B-17 and B-24 were the most common heavy four-engine bombers; but in 1944-45 a very much heavier, larger bomber, the B-29, came into use.

The B-25 Mitchell is a medium twin-engine, twin-rudder airplane.

The B-26 medium-size, twin-engine bomber was flown by WASP pilots at Gowen Field, Boise, Idaho over the eastern Oregon desert, towing target sleeves for the B-17 heavy four-engine bomber crews to practice their gunnery on a moving target in the skies.

The most famous/infamous B-29 is the Eanola Gay, which sits on the asphalt at the Paul Gerber Facility at Silver Springs, Maryland, to be stored or restored for the Smithsonian Archives in Washington, D.C.

"**C**" is the designation for cargo aircraft. The "Work Horse" C-47 Douglas, is still in use today, some fifty years later. In 1942-45, it carried people, troops and supplies all over the world. It is a very dependable, reliable airship—truly, an American "National Treasure."

"**P**" signified the pursuit airplanes and commonly called "fighters".

One thought of why the "**P**" for pursuit might have been changed to "**F**" as designating the fighter aircraft: It might have come about because some of the airplanes, as the P-38, were modified during WWII to be designated "photo recognizance" and the "P" was confusing. The P-61 was used in Photo-Recognizance also, and this twin-engine Black Widow Night Fighter was the fastest pursuit airplane in WWII. The P-59 Bell, the early jet, had a fuel "eat-up" problem which was not solved until after the war ended.

The fighters accompanied the heavy bombers over their targets. Sometimes "escort" pursuits would rendezvous with the slower, heavier bombers near the target. Sometimes the fighters would leave at later times than the bombers to "catch up" and provide cover while the bombers were accomplishing their mission and were most vulnerable.

Fighters, then labeled Pursuits were:
P-38   Lockheed Lightning built by Lockheed plant in California. The first ones were built in 1937. The P-38 "pea shooters" were single-seated and fast. (Lockheed hosted a 50th anniversary party at Universal City, California in 1987 for pilots who had flown the P-38 in World War II. Twenty-nine WASP were located who had ferried this pursuit to one of Lockheed's modification plants, or to some school were pilots were trained to fly it, or ferried it to a port of debarkation where it was sent on to combat.)
P-39   Aircobra built by the Bell factory in Niagara Falls, New York
P-40   Kittyhawk, made by Curtiss factory in Buffalo, New York
P-47   Thunderbolt, made by the Republic factory at Evansville, Indiana
P-51   Mustang built by North American factories in California and Texas
P-61   Black Widow Night Fighter, built by Northrup in California
P-63   Kingcobra built by Bell factory at Niagara Falls, New York

APPENDIX II
TERMS & ABBREVIATIONS USED IN AVIATION
(1942 - 1944)

AAF - Army Air Force (same as AAC)
AAFTD - Army Air Force Flying Training Detachment
ATA - Air Transport Auxiliary. British. Both men and women were civilians during service.
ATC - Air Transport Command.
BOQ - Bachelor Officers Quarters. A general term for the barracks on the field. WASP at all 5 bases were housed eventually on the bases.
Brig - jail
Bumped - when a previously assigned passenger is replaced by a new arrival. Desperate need for airplanes gave ferry pilots high priority in travel.
C.O. - Commanding Officer
Chandelle - a spectacular maneuver in the air — half a loop then righting the craft at the top of the loop.
CIC - Counter Intelligence
Cone of Silence - when you have passed over the emitting tower for the A and N signals. A chart in each cockpit tells where the landing field is in relation to the Cone.
CPT - Civilian Pilot Training.
Dead stick landing - no engine, no power
Dog-leg - cut across the airways path, short-cut; not recommended.
E6B - Navigation tool consisting of four plastic discs showing distance, gas consumption, air speed and other aids. (early slide-rule computer)
Empennage - British term for fuselage
ETA - Estimated time of arrival
FG - Ferrying Group
Ground loop - circle or semi-circle an airplane goes into when it gets out of control of the pilot during landing. This usually tips the airplane. On the fabric-covered planes, it damages the fabric on the under side of the wing.
HOT - Pilot or plane meant a fast, speedy person or plane.
HQ - Headquarters
ID - identification used on every base to get you in; also necessary delivering and picking up a new ship or moving an old one. This, with your orders from home base, was a "must" for traveling across the USA. A thumbprint on a telegram was sufficient; we wrote our own TR using ID, orders and thumbprint.
John - the toilet
Leg - one stop to the next; a term largely used for refueling stops
Masks - For flights above ten thousand feet altitude, oxygen masks were required. All ferrying pilots were issued face masks, and all pursuit airplanes had oxygen tanks. It was common for pilots in pursuit to pass the word that to recover from a hangover a whiff of oxygen was the best remedy!
MATS - Military Air Transport Service
OCS - Officer Candidate School
(same as OTS)
OTS - Officer's Training School. WASP believed they were to be militarized, but the women remained civilians throughout their services.
Pickle the airplane - As the planes were stacked on the decks of convoy ships at the ports of embarkation, the wings were removed and a coating was put on the metal to preserve the plane from sea water.
Roger - to acknowledge and agree
Roger, Wilco - understood and will comply
RON - Remain overnight. Symbol used in evening telegram back to home station, reporting your name, ship identification and the reason for stopping.
SBD - Navy airplane, the A-24, Douglas Dauntless. WASP at Long Beach ferried a great many of these. Mainly WASP flew them in tow target missions to train gunners to shoot at a moving target in the Air Training Command.
Scuttle - damage beyond repair
Selling a plane - after delivering to destination, turning over the ship's papers, the lose equipment that went with the aircraft, such as earphones, first aid kit and ship's log.
Shooting landings - Practice time - either

"touch and go" without stopping the aircraft, or coming to a complete stop, taxiing back to the start of the line, and heading down the runway to have another "go around". Five landings and five take-offs was standard procedure for refresher course, or renewal of skill in the delicate art of landing.
SNAFU - Situation Normal All Fouled Up. Referred to things generally that went wrong.
SOP - Standard operating procedure
Tail dragger - aircraft with skid instead of tail wheel.
TARFU - transport plane (cargo or passengers)
Tarmac - British term for paved surfaces, like runways.
TD - touchdown
TDY - Temporary duty at another station.
TO - Technical Orders. A restricted document giving all the specifics on the aircraft. These were periodically updated.
This term is also used to indicate take-off.
TR - travel request for a ticket on commercial airlines usually to home base when we could not pick up some other airplane when we delivered one.
War weary - plane that had seen much use, whether in combat or by students who were rough on the controls, especially on landings.
Wash out - to fail (to pass)
Woofteddies - nickname for the women cadets in Cochrans Women's Flying Training Detachment
WTS - War Training Service or sometimes called the War Training Program. (WTP)

APPENDIX III

MORSE CODE SYMBOLS
ALPHABET

A	.—	N	—.
B	—...	O	———
C	—.—.	P	.——.
D	—..	Q	——.—
E	.	R	.—.
F	..—.	S	...
G	——.	T	—
H	....	U	..—
I	..	V	...—
J	.———	W	.——
K	—.—	X	—..—
L	.—..	Y	—.——
M	——	Z	——..

NUMERALS

1	.————	6	—....
2	..———	7	——...
3	...——	8	———..
4	....—	9	————.
5	.....	0	—————

References

REFERENCES

A.A.F. Air Transport Command Ferrying Division Yearbook, Fifth Ferrying Group, Love Field, Dallas, Texas. Baton Rouge, Army and Navy Publishing Co. of Louisiana, 1944.

Aviation Quarterly, B. Kimbal Baker. Vol. 7 No. 3, pp. 239-263.

Aviation Quarterly, March 1986. Yvonne C. Pateman, "These WASP flew B-17's".

Brief Glory, the story of A.T.A., E.C. Chessman. London & Leeds, Petty and Sons Ltd., 1946, Prescott Books, second printing 1979.

Code One Magazine, General Dynamics, April 1989.

Document 55, History of the Women Pilots of the Ferrying Division, 1st version Nov 7, 1944. Captain Walter J. Marx.

Fighting Eagles, edited by Phil Hirsch, Charles Lindbergh's Heroic Role in World War II, by Van Heatherly. Pyramid Books, NY, 1961, pp. 7-29.

Final Report on the WASP program, Cochran, not dated.

For God, Country and the Thrill of It, Anne Noggle. Texas A & M University Press, College Station, 1990.

General Arnold Comes to Avenger Field, Alberta Fitzpatrick Head. West Texas Historical Assoc. Yearbook, Vol. I, 1974.

Hearing: Recognition for the Purposes of VA Benefits. Before the Committee on Veteran's Affairs, United States Senate, 95th Congress S. 247, S. 1414, S. 129 and Related Bills, May 25, 1977. US Govt. Printing Office 1977.

Hearing: The Role of Women in the Military. Before the Subcommittee on Priorities and Economy in Government of the Joint Economic Committee, United States Senate, 95th Congress, July 22 and Sept. 1, 1977. US Govt. Printing Office 1977.

Hearing: To Provide Recognition to the Women Airforce Service Pilots for their Service During World War II, Before the Committee on Veteran's Affairs, House of Representatives, 95th Congress. First session on granting veteran's status to WASP. Sept. 20, 1977. US Govt. Printing Office 1977.

History of the Air Transport Command Women Pilots in the Air Transport Command, Historical Branch, Intelligence and Security Division, Headquarters, 1946.

Jackie Cochran, The Story of the Greatest Woman Pilot in Aviation History. Maryann Bucknum Brinley. New York, Bantam 1987.

On Final Approach, Byrd Howell Granger. The Women Airforce Service Pilots of World War II. Scottsdale, AZ, Falconer Publishing Co., 1991.

On Silver Wings, Marianne Verges. New York, Ballantine Books, 1991.

Over the Hump, General William H. Tunner, reprinted by Office of Air Force History. NY, Duel, Sloan and Pearce, 1964.

Pilot Logbooks of flight times. Kay Gott, Helen Schaefer, Gretchen Gorman, Nelle Carmody.

Propeller Annie, Glen Kerfoot. The story of Helen Richey, the real first lady of the airlines. Lexington, KY. The Aviation History Roundtable, 1988.

Renascence, used by permission from the estate of Edna St. Vincent Millay.

Report of the Air Surgeon's Office on Wasp Personnel, 1945. Captain Nels O. Monserud. From the files of the Air Force Museum, W-P AFB, Ohio Research Dept. File L2.

Sisters in the Sky, Adela Riek Scharr. Vol. I - The WAFS & Vol. 2 - The WASP. Patrice Press, 1986.

Taped Interviews
>Betty-Jane Bachman, Marie Muccie, Kay Gott, Dallas, TX, June 20, 1972.
>Ruth Daily , Kay Gott, Tucscon, AZ, Feb. 16, 1987.
>Betty Shea, Kay Gott, Alameda, CA, June 8, 1987.
>Robert Sandoz Leveaux, Kay Gott, Tuscon, AZ, Feb. 17, 1987.
>Dorothy Kocher, Kay Gott, Tacoma, WA, Oct. 22, 1988.
>Iris Cummings, Catherine Vail, Kay Gott, Alta Loma, CA , May 18. 91987.
>Vi Thurn, Betty Scantland, Kay Gott, Asilomar, CA, April 16, 1981.
>Nancy Baker, Kay Gott, Charleston, SC, Sept, 20, 1988.
>Gayle Bevis, Kay Gott, Charleston, SC, Oct. 1, 1988.

The Eagle in The Egg, Oliver La Farge. Houghton Mifflin Co., The Riverside Press, Cambridge, Boston 1949.

The Forgotten Pilots, Lettice Curtis. A story of the Air Transport Auxiliary 1939-1945. Great Britian, The Eastern Press Ltd., London and Reading, 1982, Second Edition.

The Log Book, 319th AAFFTD, Houston Texas. Yearbook of Army Air Force Flying Training Detachment Women Pilots of Classes 1943. 43-W-1, 43-W-2, 43-W-3.

The Putt-Putt Air Force, The Story of the Civilian Pilot Training Program and the War Training Service 1939-1944, Patricia Strickland. Department of Transportation, Federal Aviation Administration, Aviation Education.

The Small College Talks Back, Dr. William W. Hall, Jr. New York, Richard R. Smith Publishers, Inc. 1951.

The Stars at Noon, Jacqueline Cochran. Atlantic Monthly Press Book, Little, Brown and Company, Boston 1954.

Those Wonderful Women in Their Flying Machines, Sally Van Wagenen Keil. The Unknown Heroines of World War II. New York, Rawson, Wade Publishers, Inc. 1979.

United States Women in Aviation 1940-1985, Deborah G. Douglas. Washington & London, Smithsonian Institution Press, 1991.

United States Women in Aviation, Claudia M. Oakes. Smithsonian Inst. Press, Washington D.C., 1985.

WASP Rosters.

WASP Songbook. Compiled and presented by Class 44-W-10.

West With The Night, Beryl Markham. San Francisco, North Point Press, 1983.

What Comes of Training Women for War, Dorothy Schaffter, American Council on Education, Washington, Oct. 6, 1948.

Women Aloft, Valerie Moolman. Alexandria, Virginia, Time-Life Books, 1981.

Women of the Air, Judy Lomax. New York, Dodd, Mead and Co. 1987.

Women Pilots in the Air Transport Command, Oct. 1942-Dec. 1944. Hq. Air Transport Command, Statistical Control Division, March 1945. Mimeograph source unknown.

Women Pilots of World War II, Jean Hascall Cole. Salt Lake City, University of Utah Press, 1992.

Writing Women's Lives, Carolyn Heilbrun, Ballentine Books, NY. 1988, 1st edition.

Index

2nd Ferrying Group, Wilmington, Delaware 95, 96, 100-102, 104, 107, 112, *114*, 130, 171
3rd Ferrying Group, Romulus, Michigan 101, 104, 107, 119, 120, 121, 125, 134, *137*, 176, 209
4th Ferrying Group, Memphis, Tennessee 61
5th Ferrying Group, Dallas, Texas 139, 140, *160*, *162*, *166*, 200, 204, 209
6th Ferrying Group, Long Beach, California 101, 125, 167-169, 171, 173-175, 177
7th Ferrying Group, Great Falls, Montana 36, 61
21st Ferrying Group, Palm Springs, California 176, 197-199, 202, *208*
33rd Ferrying Group, Kansas City, Kansas 100

A

A-20 64, 171, 174, 179, 180, 190, 200
A-20A *65*
A-24 Douglas Dauntless (SBD) *45*, *180*, 184
A-25 *179*
AAFFTD, Houston, Texas 37
Abell, Helen T. *91*, 140, *160*, *162*, *163*, *166*
Adams, Ruth 14, *89*, 96, *113*, 113-115
Aero Med Laboratory 42
Air Transport Auxiliary (ATA) 23-25
Air Transport Command (ATC) 27, 35-37, 43, 44, 46, 60, 62, 65, 66, 72, 152, 169-171, 175, 209
Albuquerque, New Mexico 98
Alcatraz 152
Alexander, Eleanor A. *91*, 198, 206
Alleman, Virginia A. "Ginny" 96, *80*, *104*, *111*, 161
Allen, Myrtle 25
Amarillo, Texas 142
American Airlines E6B 148
American Airlines 151
Anderson, Ruth E. *85*, 96, *102*, *105*, *116*, 149
Archibald, Betty J. *83*, 120, *126*, *128*, *136*
Armstrong, Merle *20*
Army Air Corps 62, 151
Army Air Force (AAF) 21, 46, 151
Army Air Force Ferrying Division Domestic Wing 27
Army Air Force Officers 71
Army Air Force School of Applied Tactics 147
Arnold, General Hap 13, 34, 72, *169*, *171*
Ascension Island 60
AT-6 *60*, *99*, 101, 102, 110, 111, 134, *153*, 161, *172*, *178*, *179*, 183, *187*, 192
Atlanta, Georgia 69
Avenger Field, Sweetwater, Texas 25, *114*, 134

B

B-17 60, 141, 150, 174, 190
B-18 Bomber *21*, 45, 144
B-24 60
B-25 (Mitchell Bomber) 64, 100, 101, 171, 174, 178-180, 190, 202
B-25A (North American Mitchell Bomber) *204*
B-26 (Martin Marauder) 64, 101, 150, 204, 209
B-29 45
Bachman, Betty Jane "BJ" 13, *38*, *77*, *78*, 140, *160*, 160-162, 200
Baker, Nancy Lee *91*, *96*, *104*, 107-110, *114*, 134
Barrick, Helen F. *92*, 120, *128*, 130, *133*, *136*
Basic Trainers 44
Batson, Nancy 26, 28, 29, 42, 65, 68, 79, 96, 100, *102*, *104*, 105, *114*, 115
Batten, Bernice *43*, 44
Bayshore Hotel 97
BC-1 111
Beard, Marianne I. *88*, *96*, 105
Beaty, Lauretta 65, *83*, *104*, 134, 168, *177*, *185*, *193*, *196*
Beechcraft *43*, *192*
Belem, Brazil 60
Bell Factory, Niagara Falls, New York 61, 62, 66-68, 123, 125, 148, 150, 155-157, 175-177, *183*, *185*, 199, 204
Bellasai, Marcia Courtney 104
Bernheim, Kathryn "Sis" *26*, *28*, *86*, *96*, *104*, 112, 149
Bevis, Gayle D. 140, *155*, *156*, *157*, *160*
Birge, Grace Pitkin *84*, 101, 107, 120, *135*
Bismarck, North Dakota 123, 126, 176
Blakely, Ronald *19*, *20*
Boeing 43
Bohn, Delphine *26*, *28*, 35, 65, *140*, *141*, *162*, *166*, 173
Boise, Idaho 192
Bolen, B. P. 178
Bolish, Juanita *83*, 120, 121, 125, *126*, *136*, *137*, 176
Bombers 33
Bothwick, Jean 104
Bowden, Mary Louise *93*, 120, 130-132, 134, *136*
Boyd, Rene 67
Braniff 172
Brinley, Maryann Bucknum 33
British ATA 100, 105
Brooks, Hazel *166*
Brown, Genevieve *84*
Brown, Mason *19*
Brownsville, Texas Pursuit School 58, 100, 101, 103, 105, 112, 113, 115, 121, 125, 130, 132, 134, 155, 175, 176, 178, 184, 186, 188, 190, 199, 204
Bryan, Texas 110, 157
BT-13 (Vultee) *43*, 44, 98, 148, 179, 182
BT-15 (Vultee) 98

Buffalo, New York (Curtiss Factory) 68, 100, 123, 146, 175
Burchfield, Phyllis 26

C

C-47 (Douglas Cargo Skytrain) 103, 144, *145*, 156, 170, 171, 174, 188, 204
C-49 103
C-60 68-69, 97
Cadman, Vivian C. *89*, 198, *203*, *204*, 206, 208
Caldwell, Idaho 141
Caldwell, Mildred "Duke" 104
Calhoun, Peggy 196
Callaghan, Claire G. *81*, 120, *127*, *135*, *136*
Caracas, Venezuela 60
Carlstrom, Marion V. 65, *83*, 140, *153*, *160*, *162*, *163*, *166*
Carmody, Nelle L. *91*, 140, *160*, *162*, *163*, *166*, 203
Carter, Ruth 104
Cartwright, Roger *20*
Castle, Margaret L. *86*, 198, *203*, 206, 208
Cessna Bobcat 36, *124*
Chadwick, Pat *136*, *137*
Chapin, Emily 24, 25
Chemical Warfare Service, San Francisco 175
Chennault, General Claire L. 152
Chester, Gene *20*
Cheyenne, Wyoming 142, 148, 202
Chicago, Illinois 130, 189
China-Burma 70
Churchill, Prime Minister Winston 23
CIC (Counter Intelligence) 142, 143
Cincinnati, Ohio 170, 171, 182
Civil Aeronautics Authority 18
Civil Air Guard 23
Civil Air Patrol 143
Civilian Pilot Training program (CPT) 18-19, 21
Clair, Virginia "Tex" 66, *82*, *96*, 107, *116*
Clapp, Ernest *20*
Clark, Grace 136
Clark, Harry 19
Clark, Helen Mary 28, 35, 61, 65, 66, *70*, *71*, 96-98, 100, 112, *114*
Clarke, Susan Parker 210
Class I 62, 180
Class II 35, 180
Class III 62, 101, 180
Class IV 64, 65, 171, 180, 183, 190, 202, 204
Clay, Captain Jack 115
Clovis, New Mexico 209
Cochise, Texas 44
Cochran, Jacqueline 10, 14, 15, 24, 25, 27, *32*, 33-35, 37, 40, 45, *103*, 105, 123, 143, 161, 169, 175, 199, 202
Colbert, Mary Lou 58, 65, *80*, 198, *199*, 204, 206, *208*
Colburn, Dorothy R. *82*, *96*, 100, *101*
Colby, Cornelia Y. "Connie" *81*, 140, *160*, *163*, *166*

Coleman, D. Lewise "Lew" 65, 77, 168, 174, 175, 185, 196
College of Idaho 17, 21
Collier's 124
Collins, Edna 43, 44
Collins, Major Eileen 210
Commodore Hotel, New York City 43
Condit, Shirley 104
Congress 72, 209, 210
Conner, Lillian M. 60, 84, 87, 101, 120, 135, 136
Cook, George 19
Coolidge, Arizona 150, 182
Coon, Joe 20
Court of Inquiry 24
Cowart, Gwendolyne E. "Gwen" 88, 96, 114, 116
Cox, Captain 113
Crinklaw, Virginia B. 90, 120, 135, 136
Cummings, Iris 65, 76, 168, 174, 179, 185, 196
Curtis, Lettice 24
Curtiss Factory, Buffalo, New York 68, 100, 123, 146, 175
Cusack, Lana Blanche 60, 86, 168, 192, 193

D

d'Ambly, Jeanne 104
Dahmes, Sylvia A. 92, 120, 127, 128, 136
Dailey, Ruth 43, 44, 65, 82, 140, 141-143, 154, 160, 162, 166
Dakar, Africa 60
Dallas, Texas (5th Ferrying Group) 37, 45, 60-62, 72, 110, 141, 142, 144-148, 151, 152, 154-157, 161, 172, 173, 188, 192, 202
Darling, Mary E. 81, 120, 124
Davis, Irene G. 93
Dayton, Ohio 142
Daytona Beach, Florida 71
DC-3 (Douglas) 101, 144, 180
Deaton, Leoti 37, 130
Deming, New Mexico 44
Denver, Colorado 142
Detroit, Michigan 126
Dias, Frances "Fran" 166, 172
Dickerson, Patricia A. 85, 120, 128, 135-137
Dion, Joe 21
Disbore, Virginia 136
Disney, Walt 183
Dolton, Ed 19
Donahue, Barbara L. "Donnie" 28, 35, 79, 120, 121, 125, 136, 137
Dougherty, Dora 45
Dressen, Ellie 196
Dunaway, Carl 20
Dunkirk 23
Dustman, Pete 20

E

Eager Beavers 57
Eames, Betty 43, 44

Earhart, Amelia 10, 17, 36, 72
Eastern Airlines 107
Edmonton, Canada 101
Edwards, Rebecca H. "Becky" 89, 198, 206, 208
El Paso, Texas 44, 113
Ellington Field, Houston, Texas 35
Ellington Post Exchange 37
Emerson, Norma Jane 65, 78, 140, 158, 162, 166, 200
Engle, Mary Edith 166
Erickson, Barbara Jane "BJ" 28, 29, 35, 58, 63, 65, 75, 98, 168, 169-171, 183, 185, 190, 196
Evansville, Indiana (Republic Factory) 61, 68, 69, 101, 102, 106, 107, 170, 171
Ewing, Gayle 166

F

F-6K (North American) 72, 144
Fairbanks, Alaska 36
Fargo, North Dakota 141, 176, 199
Farmingdale, Long Island (Republic P-47 Factory) 61, 68, 69, 97, 99, 101, 102, 103, 106, 108, 112
Farmington, New York 97
Featherhoff, Nancy L. 85, 168, 193
Federal Aviation Administration (FAA) 18
Ferrying Command 44
Fillmore, Carole 77, 168, 174, 185, 196
Fisher, Enid C. 85, 168, 191, 193
Fleishman, Lt. Alfred 37
Flynn, Anna L. 72, 87, 96, 103, 104, 111, 134, 149
Forest Lawn Cemetery 123
Fort, Cornelia 44, 210
Fort Myers, Florida 150
Fort Worth, Texas 25, 123, 145, 146
Fort Wayne, Indiana 145, 170, 171
Franckling, Ruth "Ruthie" 81, 198, 199, 200, 208
Fremd, Helen I. 83, 140, 163, 166
Frost, Jack 141
Fullerton Junior College, California 203
Fullwood, Rosa Lea 85, 140, 162, 163, 166
Fulton, Dorothy 26

G

Gardner, Libby 104
Garrett, Captain 141
General Hospital, McKinney, Texas 156
George, General Harold 27
Gery, Ellen H. 81, 120, 124, 135, 136, 202
GI Benefits 44
Gillies, Betty Huyler 14, 26, 28, 29, 35, 65, 70, 96, 97, 98, 100, 102, 110, 114
Glen, John 19
Gomes, Howard 20
Gore Field, Great Falls, Montana 36, 61
Gorman, Gretchen Johnson 67, 81, 140, 160, 162, 164, 166

Gott, Lois Kathrin "Kay" 13, 19, 20, 39, 40, 43, 44, 65, 67, 69, 72, 82, 140, 143-152, 143, 144, 149, 154, 160, 166, 202, 209
Gott, Myra 21
Gough, Theron 19, 20
Gowen Field, Boise, Idaho 149, 150
Gower, Pauline 23, 24
Grand Prairie, Texas 161, 172, 173, 192
Granger, Byrd Howell 58, 65, 82, 154, 198, 199, 204, 208
Graves, Gerald 20
Gray, Marjorie M. 35, 60, 82, 114
Great Falls, Montana (7th Ferrying Group) 36, 61, 62, 105, 123, 125, 132, 155, 176, 177, 185, 188, 199, 202, 204
Greenblatt, Evelyn 71
Greenfield, Fred 20
Greensboro, South Carolina 189
Gregory, Irene G. 70, 71, 96, 105, 116
Grohman, Rosalie L. 87, 96, 116
Gustafson, Carl 20

H

Hagerstown, Maryland 110
Hall, Dr. William W. 21
Hamilton, Lt. Col. Kelly 6, 209
Hamilton, Margaret Ann 81, 120, 123, 124, 136, 137
Hankins, Tad 20
Harbert, Gordon 19
Harrah, Jim 20
Harris, Thelma Nadine 89, 168, 185, 193
Harris, Virginia 160, 162
Harrisburg, Pennsylvania 98
Hatfield, England 23
Hensley Field, Texas 45, 62, 72, 113, 161, 192
Hiester, Emily 79, 96, 104, 114, 116, 121, 149
Higgins, Colonel Albert 151, 152, 160, 162
High Point Army Air Force Base, North Carolina 132
Hill Field, Ogden, Utah 149, 150
Hill, Virginia 65, 84, 101, 168, 182, 185, 190, 191, 194, 196
Hinckley, Robert H. 18, 169
Historic Document 55 13, 44-45
History of the ATC-Women Pilots 44
Hitler, Adolph 23
Hoagland, Jim 20
Hobson, Montana 125
Hoke, Jack 21
Holt, Elaine 196
Hooper, Mary 23
Hoopes, Jean 104
Hopkins, Dorothy "Dot" 90, 140, 162, 164, 166
Hotel Lassen, Wichita, Kansas 43
Houston, Texas (training school) 110, 122, 123, 161, 181, 190, 199, 202
Howe, Earl 19
Hrestu, Jennie 178, 180, 181
Hughes Airport 35
Hughes, Joan 24

Hunt, Bill *19*
Hunter, Celia M. *86, 96, 102, 105, 106*
Huntington Hotel 97
Hurricanes 33

I

Iwo Jima 200

J

James, Teresa D. *26, 28, 35, 69, 70, 71, 96, 98, 100, 105, 107, 108, 111, 112, 114*
Johnson, Ann *39, 162, 166*
Johnson, Dorothea 45
Johnson, Mary Catharine "Jary" *67, 79,* 104, 120, 121, *122, 123, 125, 136*
Johnson, Vega 65, *81,* 198, 203, 206, *208*
Jones, Bob *19*
Jowell, Virginia "Virgie Lee" *86, 96, 104, 107,* 108
Juarez, Mexico 44

K

Kansas City, Missouri 100, 147, 148, 157
Kary, Ross "Rossie" 65, *82, 140, 154*
Kerfoot, Glen 106
Kerr, Margaret E. *71, 81,* 120, *136,* 136
Ketcham, Marjorie Jane 67, 82, 120, *136,* 154
Kingman, Arizona 177
Kocher, Dorothy "Dottie" 65, 67, *83,* 168, 178-186, *179, 181, 182, 185, 194, 196*
Koutnik, Louis *19, 20*

L

LaGuardia Airport 107
Landis, Jean *86,* 168, 181, *185, 194, 196*
Lawler, Florence P. "Pat" *104,* 120, *111, 136*
Leaming, Kittie L. *84,* 168, 188, *194*
Ledbetter, Julie *104*
Lee, Hazel Ying "Ah Ying" *41, 92,* 120, 130, *132,* 141, 204, 210
Leveaux, Roberta Sandoz 23
Life Magazine 177
Lindbergh, Charles 17, 64, 65
Lindley, Ruth E. *85,* 140, *151, 162, 164, 166*
Link Trainers 40, *41,* 144
Lockheed Aircraft, Long Beach, California 61
Lockheed Modification Plant, Dallas, Texas 141, 142
Loft, Katherine S. "Katy" 65, *83,* 168, *171, 182, 185, 196*
Logan, Marge 71
Long Beach Army Air Field 186
Long Beach, California (6th Ferrying Group) 35, 58, 61, 66, 69, 98, 100, 112, 123, 141, 144, 170, 171, 178-180, 182, 184, 190-192, *196,* 188, 199
Long Island (Republic P-47 Factory) 61, 68, 69, 97, 99, 101, *102,* 103, 106, 108, 112
Long, Mitchell I. "Mitch" *88, 96, 104, 117*
Loop, Paula R. 210
Los Angeles, California 123, 204
Louisiana State University, Baton Rouge, Louisiana 175
Love Field, Dallas, Texas 35, 36, 41, 60, 62, 69, 113, 144, 146, 150, 157, 161, 172, 180, 182, 183, 192, 200, 203
Love, Nancy Harkness 13, 24, 27-29, 34, 65, 96-98, *114,* 169
Lovejoy, Alice E. 60, *104,* 120, *134,* 210
Luttrell, Virginia D. "Ginny" *83,* 140, *160, 162, 164, 166*

M

Madison, Isabel "Pete" *82,* 140, *151, 152, 153, 160, 162, 164, 166,* 183
Maffit, Norman *20*
Malary, V.A. *136*
Mandan, North Dakota 126
Mann, Marian G. *88, 96, 117*
Manning, Esther Rathfelder *28,* 113-115
Manning, Irene *109*
Mariner, Commander Rosemary 209
Markham, Beryl 36
Mather, John *19*
May, Alice Jean "A.J." 109, 120, *127, 128,* 133, *136,* 186
Mays, Rex 180-182
McCormick, Jill 60, *93, 96, 105, 107, 117,* 134
McCormick, Margaret *136*
McElroy, Lenore *136, 137*
McGilvery, Helen *26, 28,* 65, *79, 96, 104,* 112, *114*
McLean, Dorothy "Rusty" *104*
Meacham Airport, Fort Worth 172
Memphis, Tennessee (4th Ferrying Group) 61
Meridian, Mississippi 72, 144
Meserve, Gertrude "Gert" *26, 28, 29, 77, 96, 97, 100, 105, 108, 111, 114*
Middletown, Pennsylvania 98
Military Air Transport Service (MATS) 42
Millay, Edna St. Vincent ("Renascence") 112
Miller, Bert *136*
Miller, Florene *26, 28,* 35, 65, *70, 77, 147,* 168, *172, 173*
Miller, Maurine "Mimi" *92,* 120, *130, 136*
Miller, Sidney 65, *77, 78,* 140, *142, 160, 164, 166,* 192, 200
Mines Field, Los Angeles 189
Minneapolis, Minnesota 126, 199
Mitchell Field 156
Moffatt, Virginia C. 210
Mohrman, Jean *82*
Mollison, Amy 10
Monroe, Louisiana 174
Montreal, Canada 24, 101, 180
Mosquito 24
Moynahan, Rita Joan *82, 96, 104, 111, 117*
Mt. Rainier, Washington 150
Muccie, Marie 13
Mullens, Ruby *82*
Mullins, Virginia *82, 162, 166*
Myers, Marcie Jo "Jo" *86,* 120, *133,* 191

N

Naffz, Betty *82, 203*
NASA 210
Nelson, Esther *162, 166*
New Castle Army Air Base, Wilmington, Delaware 27, 96, 98, 113, 114
New Cumberland, Pennsylvania 58, 189
New York City, New York 108, 185
Newark, New Jersey 68, 69, 97, 101-103, 106-108, 146, 147, 150, 151, 157, 161, 175, 188, 189, 199, 204
Newman, Stanley *20*
Niagara Falls, New York (Bell Factory) 61, 62, 66-68, 123, 125, 148, 150, 155-157, 175-177, 183, 185, 199, 204
Nichols, Dorothy M. "Dottie" *82,* 123, 125, 168, 175-177, *176, 194,* 210
Nicholson, Mary Webb 25
ninety nines, inc., International Women Pilots 72
North American Factory, Long Beach 147, 148, 150, 154, 204
North American P-51 Factory, Hensley Field 161
Northwestern University, Evanston, Illinois 130

O

Oakes, Paul *20*
Oakland, California 152
Officer Training School (OTS) 71, 100, 106, 190, 204
Oklahoma City, Oklahoma 145
Ord, Nebraska 189
Orlando, Florida 71, 100, 106, 110, 147, 178, 184, 190, 204

P

P-38 (Lockheed Lightning) *29, 42,* 58, 61, *62, 63, 64-*66, 97, 98, 100, 106, 108, 112, 141, *142, 143,* 144, 155, *158, 170,* 171, 174, 175, 177, *178, 179,* 180, *181,* 182-184, 186, 189-191, 200, *201, 202,* 204
P-38F 98
P-38H (Lockheed Lightning) *204*
P-38L 202
P-39 (Bell Aircobra) 37, 58, 61, 62, *66, 67,* 100-102, 105, 121, 123, 125, 131, 132, 155, 175-177, 180, 185, *187,* 188, 190, 199
P-39Q (Bell Aircobra) 66, 176
P-40 (Curtiss Kittyhawk) *38,* 58, 68, *70,* 100-102, 105, 111, 112, 121, 122, 123, *124,* 146, 147, 152, 155, 175, 186-188, 190, 203
P-47 (Republic Thunderbolt) 37, 58, 61, 66, *68,*

69, 97, 99, *100*, 100-108, *105*, *106*, *107*, *110*, 111, 112, 122, 123, 132, 155, *159*, 161, 170-172, 175, 187, 188, 190, 200
P-47N 106
P-51 (Mustang) 58, 62, 69, 70, 72, 98, 100-102, 105, *113*, 121, 123, 131, 132, *145*, 146, 147, 154, 155, 157, 161, 171, *173*, 175, 177, 182, 184-186, 188-190, 199, 201, 204
P-51A ("Awesome Dragon") *69*, 178
P-51B 69, 170
P-51C 113, 147, 148, 151, *153*, *155*, 161
P-51D *103*, *144*, 147, 150, 154, *156-158*, 178, 204
P-59A (Bell) *149*, *200*
P-61 (Northrup Black Widow Night Fighter) 64, 98, 171, 174, *190*, *191*
P-63 (Bell Kingcobra) 61, 62, *67*, 98, 105, 121, 123, *126*, *128*, 130, 132, 150, 155, *156*, *157*, 175, 176, *177*, 180, 185, 188, 190, 191, *199*, *200*, *203*, 204
Palm Springs, California (21st Ferrying Group) 58, 61, 180, 199, 201, 203
Palm Springs, California (Pursuit School) 69, 100, 105, 123, 154, 160, 161, 174, 199, 200, 204
Palwandee Airport 130
Parker, Captain Jacqueline S. 210
Patrick, Leo *19*, *20*
Patterson Field 98, 184
Patterson, Layton 20
Pearce, Sarah Elizabeth "Liz" *89*, 96, 111, 117
Pearl Harbor 24, 69, 123
Pendleton, Oregon 150
Pennsylvania Airlines 100
Perkins, Lola C. *88*, *104*, 134, 191, 198, 207, 208
Phoenix, Arizona 184
Pierce, Hazel *162*
Pilots Handling Notes 24
Pilsadski, Jaiwiga 23
Pinkley, Avanell *91*
Piper Aircraft, Lockhaven, Pennsylvania 134, 177
Piper J-3 Cub *18*
Piper, William T. *103*
Pittsburgh, Pennsylvania 102
Pitz, Josephine A. "Jo" *86*, 96, 97, *102*, *104-106*
Poole, Esther D. *84*, 96, *101*, *102*, *104*, *105*
Poppaw, Gene *19*, *20*
Port Isabel 186
Porterfield (a plane) 143
Portland, Oregon 126, 148, 150, 177
PQ-8 (Culver) 45, *130*, 209
PQ-14 209
Pracht, Hazel W. *83*
Prince Hotel, Bismarck, North Dakota 125
Propeller Annie 106
PT-19 (Fairchild) *26*, 98, 110
PT-26 101

R

RA-24B 101
Raines, Hazel 25
Ramsey, Nadine B. 65, *81*, 168, 177, *178*, *185*, *194*, *196*
Randolph Field, Texas 209
Red-lines 72
Reed, Bill *19*
Republic Aviation Corporation 61, 103
Republic P-47 Factory, Farmingdale, Long Island 61, 68, 69, 97, 99, 101, *102*, 103, *106*, 108, 112
Richards, Helen 28, 65, *77*, 100, 168, *185*, *192*
Richey, Helen 13, 14, *25*, *71*, 96, 100, 105, 106, 108
Richmond, Henrietta "Hank" 178
Rife, Wayne *19*
Roberts, A. *20*
Roberts, Warren *20*
Robertson, Jean *196*
Rock Springs, Wyoming 148, 149, 150
Roden, Amos *19*
Romulus, Michigan (3rd Ferrying Group) 35, 36, 60, 62, 66, 69, *104*, 121-123, 125, 126
Romulus, Michigan 36
Roosevelt, Eleanor 126
Roosevelt, Franklin D. 18
Rosso, Mary M. *86*, 96, *102*, *105*, *117*
Royal Air Force (RAF) 23
Ruddell, Gerald *20*
Russell, Barbara 65, 198, 207
Rutherford, Madge A. *85*, 168, 186-189, *186*, *187*, *195*
Rutherford, Sherm 187, 188
Rwett, Bruce *20*

S

Sacramento, California 209
Salt Lake City, Utah 144, 145, 151
San Antonio, Texas 209
San Pedro, California 170, 171, 180
Sanders, Bill *20*
Sandoz, Roberta 24
Saturday Evening Post 42
Scantland, Mary E. "Betty" *87*, *104*, 120
Scarsdale, New York 134
Schaefer, Helen M. *90*, 140, 141, *159*, *160*, *162*, *166*
Schaffler, Dr. Dorothy 12
Scharr, Adela R. 28, *29*, 35, 121, 168
Scott, Captain Robert E. 70
Scott, Dorothy F. 28, *77*, 100, *138*, 140, 160, 210
Scott Field, St. Louis, Missouri 98, 145
Scott, Jane S. *92*, 120, *133*, *136*
Seattle, Washington 169
Sharp, Evelyn G. 28, *58*, 65, 141, 168, *169*, 180, *189*, 210
Shea, Elizabeth C. "Betty" 59, 62, *82*, 120, *125*, *136*, 176, 209

Shields, Anne *104*
Skinner, Ernest *20*
Smitty *130*
Snyder, Frances M. "Fran" *92*, 120, *125*, *136*
Spitfire 23, 24, 33
St. Johns, New Brunswick 24
St. Joseph, Missouri (Instrument School) 62, 100, 101, 103, 171, 178, 179, 183, 184, 190
St. Louis, Missouri 98
St. Paul, Minnesota 126, 199
Starns, Cliff *20*
Stearns, Thornton *20*
Steiner, Isabel M. *84*
Stinson (hospital ships) 101
Straughan, Jane S. *79*, 96, *99*, *104*, *114*
Strother, Dora Dougherty 211
Swan Island Base, Portland, Oregon 150
Sweet, Virginia L. 60, *93*, 120, *127*, *137*
Sweetwater, Texas 100-103, 105, 106, 110, 130, 134, 169, 180, 181, 183, 186, 190, 203, 211

T

Tackaberry, Betty C. *81*, 168, *185*, *195*, *196*
Tacke, Magda *162*, *166*
TARFU 42, 69
Tate, Janice R. "Jan" *83*, 140, *160*, *165*, *166*
Taylor, Jean *162*, *166*
Taylor, Margery 39, *90*, *162*, *166*
The Ferrying Service 34
The Fifinella Gazette 33
The Piper Girls 103
The Sun is My Undoing 126
Thomas, Leland "Lee" *19*, *20*
Thompson, Eleanor E. *84*
Thompson, Katherine Rawls 26
Thompson, Ruth R. 65, *76*, 168, *185*, *196*
Thurn, Violet C. "Vi" 14, 70, 140, *155*, *160*, *166*
Tiger Moths 23
Tinsley, Jean Kaye 210
Tomlinson, Evelyn L. *83*, 140, *151*, *160*, *162*, *165*, *166*
Tompkins, Gertrude V. "Tommy" *91*, 140, *165*, 210
Towne, Barbara 28, 58, *75*
Training Command 44, 62
Trammell, Evelyn Lacy *87*, 168, *185*, *195*, *196*
Trebing, Elizabeth M. "Met" 210
Trebtoske, Joan M. *92*, 120, *121*, *122*, *136*
Trees, Ruth Grimm *80*, 96, *158*
Trench, Jean *83*, *104*
Trick, Captain Carl 153
Trotman, Mary Tufts 65, *80*, 198, 207
Truax, Edith Deborah "Debbie" "Deb" *87*, 168, *185*, *195*, *196*
Tucker, Marjorie *20*
Tucson, Arizona 44
Tulsa, Oklahoma 98, 145, 199
Tunner, Colonel William 27
Tunner, General William H. 36, 42, 60, 66

Turmes, Franklin *19, 20*
Turner, Helen A. *86,* 140, *157, 158, 166*

U

UC-78 (Bobcat) 35, 45, 115
UCLA 175
Uinta Mountains, Utah 145
Urban, Harriet *82*
Utica 156

V

Vail, Catherine "Cappy" 65, *78,* 198, 200-202, *200, 201, 208*
Van Doozer, Katherine 23, 25
Van Nuys, California 98, 123, 175
VE Day 144
VJ Day 144

W

Wagenseil, Martha D. 65, *81,* 176, 198, *205,* 207, *208*
Walker, Lucy *162, 166*
Walker, Major *104*
War Training Program (WTP) 21
Ward, Barbara 181, 210
Ward, Bill *19*
Warner Bros. 177
Webb, Dorothy E. "Dottie" *85,* 168, *185, 191, 195, 196*
Wells, Stephanie 210
Wendelin, Margaret *90*
White Rock Lake, Texas 173
Whitlow, Elizabeth "Betty" "Bee Whit" *38, 78, 124,* 140, *153, 165,* 200
Wichita, Kansas 44
Wichita Falls, Texas 179
Wilkes, Rena D'Arcy "Billie" 168, *185, 195, 196*
Williams, Esther 177
Willis, Barbara *81,* 140, *162, 165, 166*
Wilmington, Delaware (2nd Ferrying Group) 35, 60, 61, 66, 68, 69, 97, 100, 102, 104-106, 108, 110, 112, 190
Wilson, Mary C. "Sleepy" *84,* 96, *102,* 103, *105*
Wink, Texas 41
Women Airforce Service Pilots (WASP) 27, 33 (inception), 46, 72 (deactivation)
Women Military Aviators (WMA) 209, 210
Women's Army Corps (WAC) 27, 37
Women's Auxiliary Ferrying Squadron (WAFS) 27, 35, 37, 44, 105, 112, 113, 141, 169, 189, 210
Women's Flying Training Detachment School, Houston, Texas 25, 33, 40, 143
Women's Flying Training Program 33
Wooster, Shorty 185
Wright Field, Dayton, Ohio 70, 161
Wright-Patterson AFB 210

X

X-rated aircraft 72

Y

Yakima, Washington 126
Young, John *20*
Younglove, Cy *170*
Yuma, Arizona 209

Z

Zuchowski, Janet J. *83,* 96, *104, 111, 118, 128*

PC in 2/95 - 5/95 out
BKM 6/95 - 9/95
MZ 10/95 - 2/96
GA 2/96 - 6/96
BC 6/96 -

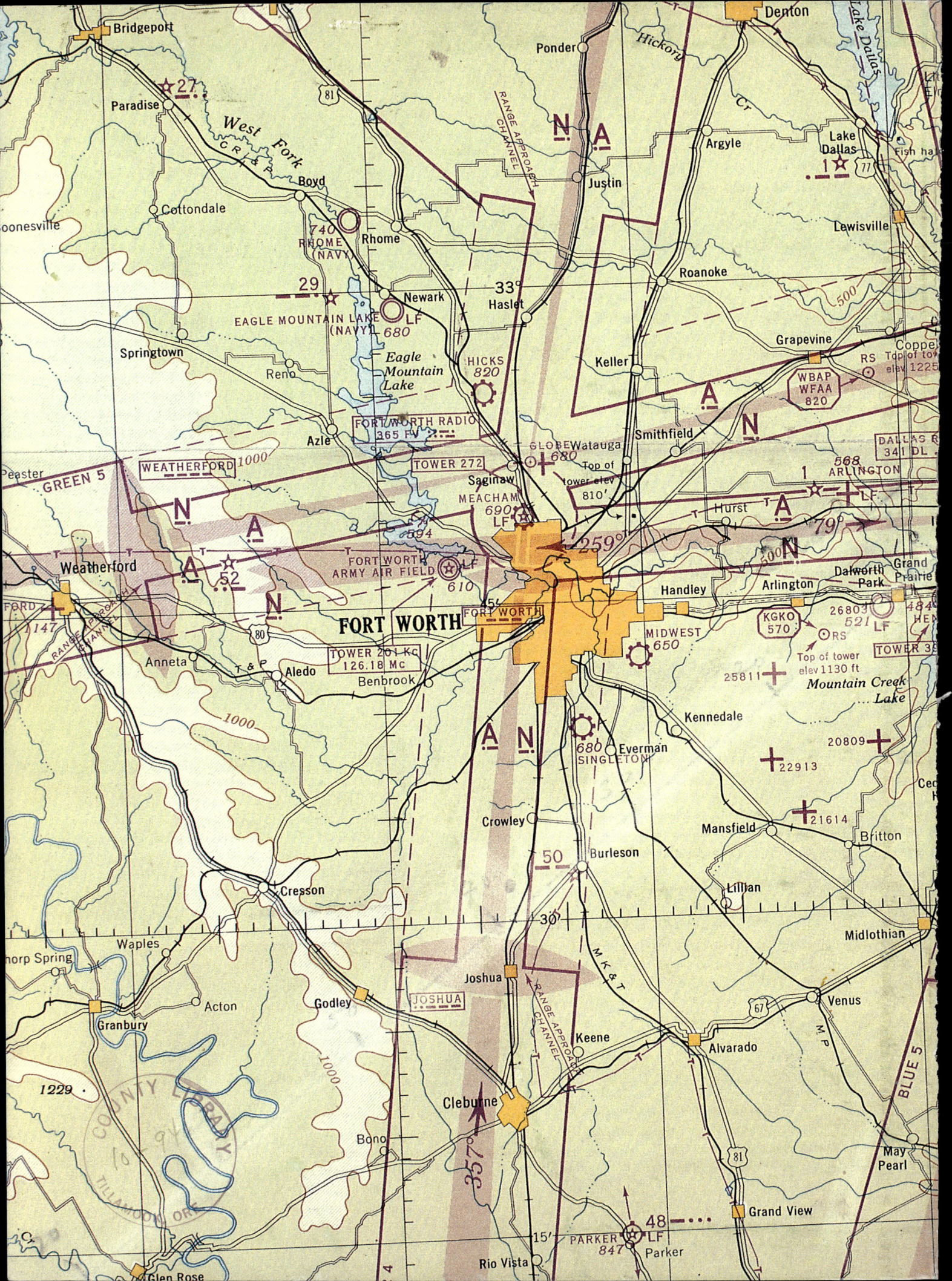